THE IMAGES OF TIME

The Images of Time

An Essay on Temporal Representation

ROBIN LE POIDEVIN

OXFORD

UNIVERSITY PRESS

OXFORD
UNIVERSITY PRESS

Great Clarendon Street, Oxford OX2 6DP

Oxford University Press is a department of the University of Oxford.
It furthers the University's objective of excellence in research, scholarship,
and education by publishing worldwide in

Oxford New York

Auckland Cape Town Dar es Salaam Hong Kong Karachi
Kuala Lumpur Madrid Melbourne Mexico City Nairobi
New Delhi Shanghai Taipei Toronto

With offices in

Argentina Austria Brazil Chile Czech Republic France Greece
Guatemala Hungary Italy Japan Poland Portugal Singapore
South Korea Switzerland Thailand Turkey Ukraine Vietnam

Oxford is a registered trade mark of Oxford University Press
in the UK and in certain other countries

Published in the United States
by Oxford University Press Inc., New York

British Library Cataloguing in Publication Data

Data available

Library of Congress Cataloging in Publication Data

Data available

Typeset by Laserwords Private Limited, Chennai, India
Printed in Great Britain
on acid-free paper by
Biddles Ltd, King's Lynn, Norfolk

ISBN 978–0–19–926589–3

1 3 5 7 9 10 8 6 4 2

For Hugh Mellor

Preface

But pardon, gentles all,
The flat unraised spirits that hath dar'd
On this unworthy scaffold to bring forth
So great an object

William Shakespeare, *Henry V*

In February 1986, the Barbican Art Gallery in London put on an exhibition entitled *Art and Time: Looking at the Fourth Dimension*. It consisted of 150 predominantly twentieth-century artworks, each representing or alluding to time, with varying degrees of explicitness. It was an intriguing and varied collection, including Umberto Boccioni's 1913 bronze figure, *Unique Forms of Continuity in Space*, René Magritte's 1928 *Les Reflets du temps*, Salvador Dali's 1933 *The Triangular Hour*, Man Ray's *Object à Détruire* (or rather his 1959 reconstruction, after the 1923 original), Etienne Marey's 1887 sculpture *Vol du goéland*, and five series of photographs from Eadweard Muybridge's 1887 *Animal Locomotion*. It also included more recent, and perhaps rather more self-conscious works, such as Rebecca Horn's 1984 *Pendulum*, which was just that, with the unexpected variation that the object suspended was a goose egg rather than the more usual brass disc. Hanne Darboven's 1971 *Ein Jahrhundert* was one of the bulkier exhibits, consisting of 365 volumes, each of 100 pages and containing a list of dates. Volume 1 contained every '1st January' in the century, volume 2 every '2nd January', and so on.

The exhibition was accompanied by a booklet, which contained a similarly eclectic collection of essays, on various aspects of time and artistic representation. It explained that one of the aims of the exhibition was to challenge the well-known distinction, expressed in G. E. Lessing's *Laocoon* (1766), between the 'arts of time' and the 'arts of space'. Paintings and sculptures, such as those exhibited here, were counted by Lessing as among the arts of space: that is, they were capable of representing only spatial properties, not temporal ones. What many of the exhibits appeared to demonstrate, however, was the variety of ways in which time can be successfully represented spatially.

When I visited this exhibition, I was a research student, writing a Ph.D. dissertation on the philosophy of time and change. Many of the artworks were of intrinsic interest, but my imagination was also caught by the general theme, and perhaps I hoped the visit might generate an idea that could be incorporated into the dissertation. It did not do so, directly at least, but it started me thinking about the nature of temporal representation, and I was also struck by one of the less opaque passages in the exhibition booklet, a quotation from an essay by Ernst Gombrich: 'While the problem of space and its representation in art has occupied the attention of art historians to an almost exaggerated degree, the corresponding problem of time and the representation of movement has been strangely neglected' (Gombrich 1964: 293). Might the pictorial representation of time raise similarly neglected philosophical problems? It was another ten years, however, before I tried to engage philosophically with Gombrich's response to Lessing's dichotomy.

Not long after this visit, I became interested in the representation of time in fiction, stimulated by the suggestion, first made in J. E. McTaggart's famous article 'The Unreality of Time' (1908), that, while we can view fictional events as occurring one after the other, we cannot make sense of any of them being past, present, or future. That did make its way into the dissertation, and was also the subject of my first full-length published article.

In 1996, another stimulus came along in the form of Christoph Hoerl's Oxford D.Phil. dissertation, 'Keeping Track of Time', for which I was the external examiner. It explored issues of what is sometimes called 'cognitive dynamics': the way in which we change our beliefs to keep track of reality through the passage of time. It drew my attention to issues of temporal representation to which I had not given much thought before, and also alerted me to John Campbell's exploration of memory in his *Past, Space and Self* (1994). Campbell's linking of the epistemology of memory with realism about the past further encouraged me to think of ways in which issues concerning the mental representation of time might interact with the kinds of issues in the metaphysics of time that I had spent the last ten or so years thinking about.

There followed in quick succession yet further stimuli to thought about temporal representation: an essay by Roger Teichmann on the impossibility of a tenseless language (1998), a typescript by David Cockburn (later published as *Other Times: Philosophical Perspectives on Past, Present and Future*), and an invitation to contribute an article to the

Stanford Online Encyclopedia of Philosophy on the perception of time. I was also given the opportunity to run a course on time and representation for MA students and third-year undergraduates at Leeds, based on some of the ideas in this book. Eventually, what started as a collection of somewhat tenuously related questions began to resolve themselves into a common theme: how we integrate temporal features of the world into our understanding of the mechanisms behind representation in different media—perception, memory, art, and fiction. At that point, it seemed worthwhile to embark on a full-length treatment of the topic.

It might perhaps be thought surprising that I have said very little about emotion in this study, in view of the very extensive debate over Arthur Prior's (1959) 'Thank Goodness That's Over' argument against the original form of the B-theory of time. This argument has since been expanded and developed as a way of showing that states of relief, anxiety, etc., are only rational if time is viewed as an A-series. But precisely because there has been so much debate, to which I have nothing to add, and also because in Chapter 2 I characterize emotional states that have representational content in terms of other psychological states, I have omitted any discussion of Prior's argument here. The interested reader is referred to Oaklander and Smith (1994), Part III, and Cockburn (1997), Part I.

In this study I have drawn quite extensively from the following previously published articles: 'Time and the Static Image', *Philosophy*, 72 (1997), 175–88; 'Egocentric and Objective Time', *Proceedings of the Aristotelian Society*, 99 (1999), 19–36; 'Can Beliefs be Caused by their Truth-Makers?', *Analysis*, 59 (1999), 148–56; 'The Perception of Time', *Stanford Online Encyclopaedia of Philosophy* (2000), <http://plato.stanford.edu/entries/time-experience/>; 'Fate, Fiction and the Future', *Philosophical Papers*, 30 (2001), 69–92; 'Why Tenses Need Real Times', in Aleksandar Jokic and Quentin Smith, eds., *Time, Tense and Reference* (Cambridge, Mass.: MIT Press, 2003), 305–24; 'A Puzzle Concerning Time Perception', *Synthese*, 142 (2004), 109–42.

I have benefited from conversations and correspondence with many people during my pilgrim's progress along this, at times, rather steep and winding road, especially Craig Bourne, John Campbell, David Cockburn, John Divers, Christoph Hoerl, Jonathan Lowe, Andrew McGonigal, Joseph Melia, Adrian Moore, Nathan Oaklander, Sean Power, Rebecca Roache (who also kindly corrected a complete draft), Jon Robson, Katia Saporiti, Peter Simons, Quentin Smith, Jonathan Tallant and Roger Teichmann. I would also particularly like to thank

Peter Momtchiloff, of Oxford University Press, for encouraging me to write the book and to persist with it when it seemed it would never see the light of day; two anonymous referees for the Press for their detailed and immensely helpful comments; the students who took my 'Time and Representation' course between 1996 and 2003; the Arts and Humanities Research Board, who funded my research during 2001–2; Craig Callender, who kindly showed me an unpublished paper that drew my attention to the significance of the flash-lag effect for the B-theory; audiences at Bristol, Glasgow, Lampeter, Leeds, London, Kirchberg, and Oxford, for reactions to early versions of chapters of this book; Gregory Currie, whose work on fiction and pictorial representation, and reflections on the idea of a literary philosophy of time, made the writing of Part III of this book possible; Jeremy Butterfield, who showed me the importance of issues of perception for the debate between the A-theory and the B-theory; and finally Hugh Mellor, whose patient supervision of my early research in time first set me on this road, and whose work in metaphysics has been a constant source of inspiration. I dedicate this book to him.

Here, then, is the conclusion of a train of thought that began with that visit to the Barbican Art Gallery more than twenty years ago.

R Le P

December 2006

Acknowledgement

Excerpt from 'Reference Back' from COLLECTED POEMS by Philip Larkin. Copyright © 1988, 2003 by the Estate of Philip Larkin. Reprinted by permission of Farrar, Straus and Giroux, LLC.

Contents

It is also worth investigating how time is related to the soul.

Aristotle, *Physics*

Truly, though our element is time,
We are not suited to the long perspectives
Open at each instant of our lives.
They link us to our losses: worse,
They show us what we have as it once was,
Blindingly undiminished, just as though
By acting differently we could have kept it so.

Philip Larkin, 'Reference Back'

PART I

ASPECTS OF TIME AND REPRESENTATION

1
The Project

A man that looks on glass,
On it may stay his eye;
Or if he pleaseth, through it pass,
And then the heaven espy.

George Herbert, 'Teach me,
my God and King'

Late for the train, you run from the ticket office, pushing through maddeningly slow-moving crowds, and drag an impossibly heavy suitcase on to the escalator, where, blocked by the commuters in front of you, you try to distract yourself by glancing at the posters passing by. One of them is advertising an exhibition of early twentieth-century futurist art, and reproduces a famous painting of a dog trotting along on a leash. Reaching the bottom, you anxiously search the Departures board for the right platform, vaguely aware that some services (though not yours) are delayed. At last you reach the platform, where the train awaits, still motionless. Glancing at the clock, you see the second hand move to the 12 o'clock position, and, as you frantically approach the doors, with strangely leaden feet, the guard raises his flag and simultaneously blows his whistle. A few seconds later, the train departs, gathering momentum as you watch, defeated, its diminishing form. Somewhere, a bell is ringing, and as it increases in volume, filling your consciousness, you wake up, switch off the alarm, and realize with relief that the train you have just missed in your dream is in reality not due to depart for another five hours. Later, you narrate the events of the dream to a characteristically unresponsive audience over the family breakfast.

Within this vignette is a multitude of ways in which we represent time: in perception, memory, belief, emotion, art, and narrative. Throughout, each experience presents itself to you as *present*: first the purchase of the ticket, then the journey on the escalator, then the dash for the train. You believe that the train *will* depart *shortly*. Looking at the poster, you recall *having seen* the painting before. The painting itself represents *different moments* of an event. You perceive the *motion* of the second hand on the clock, the *simultaneity* of the flag waving and the blowing of the whistle, the *duration* of the sound, the fact that the train leaves shortly *after*, and its initial *acceleration*. On waking, you feel relief that the train's departure is still *future*. All these facts are themselves represented in the dream and subsequent narration.

Philosophers have long been exercised by puzzles of mental represent-ation: What makes something the object of a perception, belief, desire, emotion, or intention? How is mental representation linked to linguistic and pictorial representation? How do words refer, sentences have mean-ing, and pictures depict? But even when we have, at least provisional, answers to these questions, there remains a further set of puzzles as to how these forms of representation manage to capture aspects of *time* (and indeed space). For example, causal theories of perception hold that the perceptual representation of an object consists in that object's being causally responsible, in the appropriate way, for the character of the perception. And this seems plausible enough in the case of perception of trees, flowers, people, and the other concrete items that causal theorists of perception tend to talk about. But how is it supposed to apply to rela-tional properties such as simultaneity, precedence, and duration, which, arguably, are not themselves causal? To take another example, according to the indirect account of episodic memory (that is, a memory of expe-riencing some event), a memory tracks a past event only by virtue of being appropriately related to an original experience. But if the original experience represented an event as *present*, and the memory represents that same event as *past*, how exactly are we to understand the epistemic and semantic relationship between the earlier experience and the later memory? Yet another example: it is plausible to suppose that a realist painting counts as a depiction of a well-known rural scene by virtue of an intended correspondence between the spatial distribution of shapes and colours on the canvas and the spatial arrangement of the depicted objects. But, given that these features of the painting are entirely static, how do we account for the depiction of change and motion? A final example: the events and characters of most fictions are isolated from us—we are not (typically, but with the obvious exception of interactive fictions) part of the fiction. Now this seems entirely consistent with the idea of fictional events as exhibiting a temporal order. But is it consistent with a fictional event's being (fictionally) past, present, or future? We might appeal to a device in literary theory: the idea of a fictional narrator, from whose temporal perspective the events are being narrated. But can this account accommodate the possibility of there being fictions about parallel time streams?

One aim of this study, then, is to investigate whether, and how, otherwise plausible theories of representation, in its various forms, can be reconciled with our ability to represent aspects of time. Another aim is to discover something about the nature of time itself by looking at the nature of its mental representation. Now, according to some ways of thinking, this second aim is unrealizable. Narcissus looked into a pool and saw a face of such beauty that he fell in love with it. But every time he tried to touch it, it disappeared. What he saw was only his own reflection. What else would we expect to see, in looking at a representation of the world, but that representation itself, rather than something beyond? To study a representation is precisely not to study the object represented (at least, not directly). It is true that representation is a relational phenomenon: there is, on the one hand, the *representans* and on the other, the *representandum*. But the traditional empiricist notion that the relation between the two is one of resemblance is not a satisfactory one. (Even in the case of pictorial representation, where the resemblance account seems most promising, the relationship between *representans* and *representandum* is more complex.) Representation is rarely, if ever, transparent: one cannot simply read off the nature of the world from the intrinsic features of the representation. And if not, how is it possible for a study of representation to tell us about the structure of reality?

I am not raising a sceptical point here. The question is not, how do we ever know that our representations of the world are correct? This project begins with the assumption that many of our representations are in fact accurate, and that we can know them to be so. The question I want to pursue is, rather, what ordinary representations of time (the presentness of experience, the belief that something is about to happen, the memory of something else happening, the perception of simultaneity and precedence) tell us about the metaphysical structure of time itself. And the answer that we might initially be inclined to give to this question is: very little. For metaphysical theory is typically underdetermined by the ordinary data of experience. The phenomenal features of shape perception, for instance, do not seem to settle the question of whether we should be realists or nominalists about general shape properties like circularity. Likewise, the phenomenal features of object perception do not seem particularly to favour either side of the debate between those who view particulars as bundles of properties and those who insist on an underlying substratum. But what if we put the observational data about mental representation together with a philosophical understanding of the representation relation, and perhaps also a psychological account of the mechanisms underlying mental representation? Could we then derive some interesting metaphysical results concerning the nature of what is being represented? More specifically, can we learn anything about the metaphysical nature of time from looking at a combination of observational data concerning, and philosophical and psychological accounts of, the various forms of mental representation of time? My answer to that question is: yes, we can. I aim to convince the reader of that in the following pages.

Part I introduces some general issues concerning time and representation which set the scene for the discussion that follows.

We begin, in Chapter 2, 'Causal Theories of Representation', with three questions about representation of any kind: What is a representation? What determines its content? What determines its epistemic status? It is difficult to find an answer to the first that is both informative and general; but a central feature of representations is that their content, what they are *about*, is linked to what would make them true. (This characterization has to be adapted for those representations, such as of desire and intention, where the notion of truth does not appear to be directly applicable.) A plausible and influential answer to the second question, concerning content, is a causal one: representations are linked to what they represent causally. The third question, concerning epistemic status, asks what it is for a true mental representation of the world to be sufficiently well grounded as to count as an item of knowledge. Again, we are not concerned with global sceptical issues about whether in fact any of our representations count as knowledge, but with the question that arises once we help ourselves to the assumption that many of them do: what is the difference between those representations that count as knowledge and those that do not? Again, the most plausible and influential answer to this question is a causal one. Part of the reason why it is so plausible is that the basic form of mental representation, perception, is a causal process, and the processes by which perception gives rise to other forms of representation, namely belief and memory, are themselves evidently causal. The central idea of causal theories of knowledge is what I shall call the 'Causal Truth-Maker Principle' (or some variant thereof): namely, that those perceptual beliefs that qualify for the title 'knowledge' are caused by their truth-makers. This epistemological principle is, I suggest, a significant metaphysical tool, for it requires that the world contain items that are capable of playing two roles: those of causing and of making true our beliefs. This may seem inoffensive enough, but, as we shall see in subsequent chapters, it raises acute difficulties for our understanding of temporal representation.

Chapter 3, 'Egocentric and Objective Representation', introduces a distinction between ways of representing the location of things in the world. *Egocentric* representation is subject-centred, and reflects the subject's position in some dimension. *Objective* representation is not subject-centred, and represents the placing and orientation of things independently of the subject's position in that dimension. To put it another way, egocentric representation is *perspectival* in a way in which objective representation is not. The very distinction between these two allows us to define a concept that is central to metaphysics: the concept of mind-dependence. We can say that a certain aspect of the world is *mind-dependent* when it features in egocentric representations, but not in objective representations. This is not the only sense of mind-dependence that we can frame (and indeed, it does not capture all mind-dependent properties); but it is one of the simplest, and it is also the one most pertinent to the metaphysics of space and time. My belief that it is hot in *here*, and that the orchid is *over there* and *further away* than the Venus fly-trap, is very evidently an egocentric representation of space (or rather of the spatial positions of objects). The distinction between 'here' and 'over there' would not be a part of any spatial representation that was independent of my own spatial position, and so is mind-dependent.

When we come to temporal representation, however, things are not so straightforward. My belief that it has *just started* to rain, and that I *left* my umbrella at home, looks like an egocentric representation, in that it reflects my own temporal position vis-à-vis the events in question. But, unlike the spatial case, it does not seem as if my own temporal perspective determines whether these events are past or future. The distinction between past, present, and future looks entirely mind-*in*dependent. Whether or not an event is present does not, intuitively, depend on the existence of someone to judge that it is so. But how can the distinction *both* reflect our temporal perspective *and* pick out an objective feature of time? The recognition that a representation is perspectival suggests access to a non-perspectival conception of whatever is being represented. Do we have access to such a conception in the case of time?

This raises the issue between the A-theory and the B-theory of time, a debate which runs through much of this book. According to the *A-theory*, the distinction between past, present, and future is *not* merely perspectival or mind-dependent. The truth of 'tensed beliefs'—i.e. those whose linguistic expression would be tensed, in the sense of indicating pastness, presentness, or futurity—depends upon some fundamental feature of time: the existence of an *A-series*, the series of positions running from the distant past to the distant future. The A-theory is an attempt to capture our ordinary belief that time passes. The *B-theory*, in contrast, denies that time passes in reality. Time's structure is much more akin to that of space. Tensed beliefs simply reflect the thinker's perspective. It follows that, in attempting to describe time as it is in itself, and the objective basis of our tensed beliefs, the B-theorist will make use of tenseless temporal expressions such as 'The lunar eclipse takes place on 3rd March, 2007', which, if true, is true at all times. For the B-theorist, these are truly objective representations of time, locating events in the *B-series*, positions in which are determined just by unchanging relations to other times or events. There is, however, a line of argument advanced by some A-theorists which proposes that there are in fact no genuinely tenseless temporal expressions, and that any expressions purporting to be so are in fact parasitic upon tensed expressions. If they are right, one cannot even coherently articulate the notion that time exists but does not pass. So part of the B-theorist's project must be to establish that we can indeed distinguish between egocentric and objective representations of time: that we really can describe time from no particular temporal perspective (although in practice this has been taken for granted by most B-theorists—an exception is Moore (1997)). This task is undertaken in the next chapter, on memory, and in Chapter 8, on fiction.

Part II brings together the epistemological principles concerning the causes and truth-makers of our beliefs with metaphysical debates concerning time.

In Chapter 4, 'Retracing the Past: Memory and Passage', we encounter questions raised by episodic memory: what is the relationship between a memory of experiencing an event and the original experience of that event? What is it about the memory that tells one that the event in question is past? In what sense does memory preserve perceptual information? The causal theory of representation defended in Chapter 2 provides part of the answer to these questions, but it also sits rather uncomfortably with what would otherwise seem the natural metaphysical account of what kinds of fact memory represents, namely the A-theory. If I truly remember meeting Emily in the Art Gallery, then my encounter with her is now past, although at the time of the original experience it was present. Since that encounter, reality, as we might put it, has moved on. The encounter was once present and is now past. So what made true my original perceptual belief ('I am now seeing Emily') and what now makes true my current memory belief ('I saw Emily then') are two quite different things. But they must nevertheless be intimately connected, for otherwise it would be quite obscure how memory preserves the information contained in the original perceptual belief, and further, how the epistemological status of the memory (its counting as a state of knowing, for instance) is determined by the epistemological status of the original belief. The A-theorist will say that the connection is a logical one: the truth of my original belief entails the truth of the later belief. But it is actually quite unclear how the A-theorist can account for this 'trans-temporal entailment', a difficulty that is analogous to the problem of truth-value links for anti-realism about the past.

The two theories of time introduced in Chapter 3 thus give rise to two different models of episodic memory: for the A-theorist, memory is a matter of keeping track of changing states of affairs; for the B-theorist, it is a matter of keeping in mind a constant state of affairs. Only the B-theory, I argue, can satisfactorily account for the epistemology of episodic memory.

Perhaps the most intuitively appealing argument for the truth of the A-theory arises from reflection on ordinary perceptual experience. Every perception we have presents its object as *present*. Moreover, we are constantly aware of the passage of time: we can actually see the sand falling through the hourglass, the sun sinking below the horizon, the second hand moving around the face of the clock. Any theory that denies the real passage of time, as the B-theory does, flies in the face of this experience. This is 'the argument from experience', which in effect moves from phenomenological premises to a metaphysical conclusion. Whether or not it is a good argument is the theme of Chapter 5, 'Projecting the Present: The Shock of the Now'.

It would generally be conceded that the experienced present cannot correspond directly to any 'objective' present, since the present of experience extends over an interval, whereas the objective present, as Augustine established, is a durationless instant. In addition, the finite speeds of light, sound, and information processing entail that what we perceive is actually the (typically very recent) past. It remains true, however, that our experience is temporally limited: we do not continue to perceive the past, and we do not perceive the future at all. Can the B-theory explain this? I argue that it can, by appeal to various facts concerning causation and information processing.

The most problematic data to interpret concern the experience of succession: seeing a bird flying overhead, hearing a musical phrase, feeling the rain on one's face. It is this kind of experience that seems most clearly to intimate the passage of time. However, the data give rise to a phenomenological paradox. On the one hand, what we perceive we perceive as present. But on the other, we perceive succession, and so different states that cannot be viewed as present together, for then they would be perceived as simultaneous. How, then, can we *perceive* succession? The Jamesian doctrine of the specious present, taken up by other writers, is of no help here, it turns out, and is in any case based on a mistake. The solution, I suggest, lies in the notion of the perception of 'pure succession'.

The chapter brings together a number of considerations concerning time perception to suggest that presentness and the passage of time are projected on to the world, rather than being passively perceived. The view is parallel to projectivism about colour.

'Time perception' may with some justification be thought a misnomer. As the psychologist J. J. Gibson once remarked, 'events are perceivable but time is not' (quoted in Pöppel 1978: 713). We do not, however, register events only, but also their time order and duration. Can this be reconciled with the causal theory of perceptual knowledge, and specifically with the Causal Truth-Maker Principle? This is the subject of Chapter 6, 'The Wider View: Precedence and Duration'. The difficulty here is that duration and temporal precedence do not appear to be the kinds of item that are locatable at specific instants; yet this is what they need to be in order to be the causes of the beliefs they make true. In addition, psychological accounts of the perception of duration and precedence appear not to ascribe any causal role to the supposed objects of those perceptions. The argument of the previous two chapters, based on the nature of temporal experience, was that presentness is mind-dependent in some sense. Does the problem raised here suggest that precedence and duration are similarly mind-dependent? Something like this line of argument appears in Augustine's famous discussion of time in the *Confessions*. But the 'Augustinian' solution to our difficulty is not satisfactory. A far better approach is to revise the causal theory of perceptual knowledge. We need to recognize a form of explanation distinct from causal explanation, but often conflated with it, namely *chronometric* explanation: explanation in terms of the temporal structure of events. The consequent revision in the epistemology of perception connects in interesting ways with the objectivist/conventionalist debate over temporal metric, and the temporal asymmetry of causation.

In Part III, we turn our attention to more advanced, or more derivative, forms of representation: art and narrative fiction. But the concerns of Part II—the metaphysical debate over presentness and passage, and the understanding of temporal experience—continue to be relevant.

In Chapter 7, 'Image and Instant: The Pictorial Representation of Time', the question is asked whether, and how, pictures are capable of depicting time, change, and motion. When the eighteenth-century critic G. E. Lessing distinguished between the arts of time and the arts of space, he placed painting with the arts of space—i.e. those incapable of directly representing change. Painting, according to Lessing, can capture only an instant of time, the challenge for the artist being to choose, from the various stages of an action or event, the moment most suggestive of what is to follow. This distinction between the two kinds of art, and the notion of an instant on which it rests, were vigorously criticized by Ernst Gombrich. For Gombrich, the specious appeal of Lessing's thesis rests on a mistaken motion of change as a series of instantaneous states. Gombrich associates this with a similarly mistaken psychological thesis: that perception of change is really a series of perceptions of instantaneous states—a thesis we might, following Bergson, call the 'cinematic' view of perception. The combination of these two erroneous theses obscures our understanding of pictorial representation.

To assess Gombrich's imaginative and challenging argument, we need to look carefully at the the logical relationships between the metaphysical, psychological, and aesthetic theses in play, and in the context of a plausible account of depiction. A sketch of such an account is offered in Chapter 7, and on the basis of it, we can judge Gombrich to have reached the right conclusion (more or less), but also to have taken the wrong route to it. That static images can depict change is compatible with both the static account of change and the cinematic view of perception.

The next chapter reverts to a central theme of Part II: the reality of the A-series. But whereas our attention was previously concerned almost exclusively with the past and the present, we now turn to the future, which adds a further dimension: namely, the determinacy of other times. Chapter 8, 'The Fictional Future', explores this issue in the context of fictional narrative. The central question is whether there is such a thing as a fictional future. That is, are there true statements about what *will* or *might* happen within the fiction (more precisely, future-tensed statements that are true-in-the-fiction)? For many A-theorists, the unreality or indeterminacy of the future provides one of the key motivations for their theory. But can we square the indeterminacy of the future with the fact that some fictions appear to represent the future as *fixed*—i.e. as there being a fact of the matter as to what will happen—such as *Oedipus Rex*? It might seem surprising that there could be any conflict here. Reality, after all, is reality, and fiction is fiction. But in fact, those accounts of truth-in-fiction that appeal to possible worlds face difficulties in accommodating this aspect of the A-theory. Others, such as those that appeal to the idea of the imagination, or of fictional narrators, struggle to account for future-tensed fictional truths. The problem is this: the truth of tensed propositions (it was/is/will be the case that *p*) concerning fictional events presupposes a perspective. But we cannot find a perspective internal to the fiction that will offer us truths about the fictional future. The only viable solution seems to be to treat fiction as tenseless: in so far as a fiction has a temporal structure, that structure is composed solely of the relations of precedence and simultaneity. Fictional time does not pass. However, we do not have to banish tense altogether from our imaginative engagement with fiction.

Finally, in Chapter 9, 'The Unity of Time and Narrative', we ask whether our view of an issue concerning time's topological structure has any interesting consequences for our understanding of fictional representation (or vice versa). According to Kant, time is essentially *unified*: the idea of parallel time series bearing no temporal relations to each other is incoherent. For Kant this is just a truth about the conceptual scheme we impose on our experience, for the world itself is not temporal. But if we believe time to exist independently of mind, is there any reason to think that our time series is unique? A number of writers have argued for the necessary unity of time on the grounds that the idea of an A-series is essential to time, and we cannot coherently talk of a past, present, or future in a time series that is not our own. It is not merely the existence of an A-series, however, that is in conflict with disunified time, but the further suggestion that all genuinely temporal propositions are irreducibly tensed, a view that was the target of Chapters 3, 4, and 8. Assuming, then, that disunified time is a metaphysical possibility, we turn to the possibility of *fictionally* disunified time—i.e. the fictional representation of different time streams. But here we encounter a difficulty. Suppose we attempt to construct a story in which there are multiple time streams. It must be a *single* fiction if it is to be a description of multiple time streams, and not just a collection of separate fictions, each with its own time. But what makes it a single fiction? Arguably, the unifying feature is the temporal relations between any given event in that fiction and any other. In other words, narrative unity presupposes temporal unity. The constraints of fictional narrative may therefore support our intuitive belief in the unity of time. But that support is strictly limited, for there is an alternative account of fictional unity which allows for fictions about parallel time series. And if we do allow fictionally disunified time, as I argue we should, this has interesting implications for the fictional narrator device discussed in the previous chapter.

What, in short, I hope to offer in the following pages is both an account of temporal representation in its various forms—perception, memory, art, and fiction—and an account of the nature of time, by showing how these two things—image and reality—interact.

2

Causal Theories of Representation

> They are impressed by Nature's hand.
>
> William Fox Talbot, 'The Pencil of Nature'

2.1 WHAT IS A REPRESENTATION?

When Socrates asks Theaetetus 'What is knowledge?', the young man's first response is to instance two very different kinds of knowledge: namely, geometrical knowledge and cobbling. But this is not what Socrates wants, for he is looking for that essential mark by virtue of which something is called knowledge, not simply a catalogue of examples (Burnyeat 1990: 146c–e). If what is being asked for here is an account of the necessary and sufficient conditions for the application of a certain concept, in terms of analytically more basic concepts, there may be some concepts for which the request is not appropriate, either because there is no identifiable common element in all instances of the concept (Wittgenstein's example of 'game' is the one most often quoted in this context (Wittgenstein 1953: §§ 66–7)), or because the concept is itself analytically basic. What, then, of the concept of representation? We can easily produce a catalogue of instances: beliefs, perceptions, memories, declarative sentences, photographs, films, paintings, and sculptures (of certain kinds), and so on. We readily assent to the suggestion that these all have something in common, that their being assigned to one category is not simply arbitrary, or conventional. But can we say anything informative about what the common element is? Or are we just stuck with the bare assertion that they all *represent* the world as being a certain way? Is 'representation' analytically basic?

Even if we suspect that representation is indeed an analytically basic concept, we do not have to remain content with a catalogue of instances, for we can draw attention to the various connections that representation has with other concepts, such as object-directedness or content: a representation is *of*, or *about* something, though not necessarily an existing something. Can we say anything informative about what defines a representation's content? Take belief. What, plausibly, define the content of a belief are the conditions under which it would be *true*. But the attempt to generalize this ('a representation is that which has truth-conditions') immediately faces the objection that there are many examples of mental states, utterances, and inscriptions that, although indisputably having representational content, are devoid of truth-value, such as desires, intentions, oaths, and commands. (And only in an extended sense of 'true' can we describe a perception as 'true'.)

Can we widen the definition to incorporate these? Here is a first attempt. In the case of these non-truth-apt but representational items, there is something that plays the role that truth does for beliefs. We could describe it in general terms teleologically: there is something that representational states aim at. There is, in each case, something whose realization means that the representation in question achieves its aim. This something then defines the content of the representation: it is what the representation is *about*. We can therefore describe this as the representation's 'success conditions'. But what success amounts to differs from case to case: in the case of belief, it is truth; in the case of desire, satisfaction; and so on. We can bring out both the variety of, and also the structural similarity between, representations in a table (see Table 2.1).

Table 2.1: Representations and success conditions

The content of a	perception		veridical
	memory		accurate
	belief	is given by	true
	desire	(a proposition	satisfied
	intention	expressing)	fulfilled
	assertion	the necessary	true
	command	and sufficient	carried out
	plan	conditions	implemented
	promise	for its being	honoured
	picture		accurate

We could therefore define a representation as something whose content is defined by its success conditions. This can be both true and useful without our having to make the dubious assumption that we have a grip on the concept of content and the various success concepts such as truth that is quite independent of a grasp of 'representation'. In the case of mental representations, this characterization might, given its teleological nature, suggest an account of representation in terms of evolutionary success: the 'success' conditions of a representational state of an organism are just those conditions which, if realized, would put the organism at an advantage in the process of natural selection by virtue of being in that state (see e.g. Papineau 1987, 1993). Equally, however, we could accept the general characterization without thereby being obliged to accept an account of content in terms of natural selection.

There are, however, a number of objections to the 'success conditions' conception of representational content.

First, tempting though it is to say that having success conditions is what defines representational states, we might be hard put to define 'success' in such a way that only representational states have success conditions, without appealing to the notion of representation. For instance, one might talk of the 'success' conditions of a disposition, these perhaps being the disposition's manifestation conditions, but not all dispositions are representational.

Second, consider the act of contemplating, or entertaining, an idea. When a thought comes into my mind—for instance, that Lavinia knows something about Digby that I don't know—I may simply entertain it without actually believing it to be true. The thought I entertain certainly has content, but it seems odd, indeed inappropriate, to describe the contemplation of such a thought as having 'success conditions'. There are no circumstances we could describe that uncontroversially capture the achievement of the purpose of such contemplation. Does the contemplation of a thought *have* a purpose, always? Or is it something that just happens?

Finally, there are a series of objections to a specifically evolutionary interpretation of the success conditions account of content, the most obvious one being that we can easily imagine circumstances in which a belief has survival value, but which diverge from the content of that belief.

If we wish to avoid controversy, it seems wise to look elsewhere for a characterization of representation. Let us go back to the earlier, hastily discarded, suggestion that *truth* is the central defining notion of representation. For reasons we have rehearsed, this immediately looks implausible: a glance at Table 2.1 tells us that not all representations have truth-value. But a couple of distinctions dispel the implausibility.

Focus for a moment just on the forms of mental representation listed above. Arguably, they all have a component in common: a mental representation that we can simply call a *thought*. The thought is a representation of a certain state of affairs. The differences between mental representational types can be characterized in terms of different *attitudes* accompanying that thought: assent, doubt (withholding assent), hope, desire, resolution, fear, and so on. An act of contemplating *p* involves simply focusing one's attention on the thought that *p*, involving neither assent nor doubt. And so on. ('The thought that *p*', note, is *not* to be identified with a belief that *p*: it is simply a mental representation with *p* as its content.)

This suggestion that all mental representation involves an attitude towards a thought is clearly closely related to the characterization of beliefs, desires, and the rest as 'propositional attitudes', implying that the attitude is towards a proposition. I have no objection to this way of talking, but the proposition has to have a mental counterpart, or else how does the attitude attach to that particular proposition? And thoughts, like propositions, are vehicles of truth and falsity. The content of a thought, then, can be defined by its truth-conditions.

But what of the observation that desires, intentions, and other non-doxastic attitudes, are not capable of being true or false? We should, I think, distinguish the question of whether it is appropriate to evaluate a given mental representation as true or false from the question of whether such a representation contains, at its core, a truth-apt component. It would clearly be inappropriate, a category mistake, to evaluate a desire as 'true', or to criticize an act of contemplating a thought as 'false'. But this does not show that what is being desired, or contemplated, is something that cannot be described as true or false. Contemplating a thought, whether that thought is true or not, does not commit one to supposing that thought true, so the fact that the thought turns out to be false does not make the contemplating of it similarly false. We might, perhaps, describe the act of contemplation as 'idle', but this is not a form of falsehood.

So mental representations, the suggestion goes, are complex: they consist of a thought plus an attitude (and we had better assume, for fear of a regress, that the attitude itself is non-representational). And other forms of representations are, plausibly, representational by virtue of being connected with mental representations, and hence with thoughts. Finally, the content of a mental representation is defined by the truth-conditions of the thought that lies at its core.

What becomes of the idea of 'success conditions'—conditions of satisfaction, fulfilment, etc.? That idea is not completely lost, but it can be seen as derivative from the idea of truth-conditions. Take a desire that p: this involves an attitude towards a thought that p. The truth-conditions of p coincide with the satisfaction conditions of the desire. This conception of 'success conditions' need have no implications whatsoever for the instrumental value of the mental representation in question. Take emotions, for instance. Not all emotions are truly representational: one may feel depressed, or happy, or irritable without feeling depressed or happy or irritable about anything in particular. But many emotions are clearly representational: you are relieved *that* the interview is over, excited *that* Cousin Timmy is coming to stay, annoyed *that* the car refuses to start. But what exactly counts as 'success' for these emotions? If we think of it in purely instrumental terms, as preparing us in appropriate ways for what is to come, we will not always capture the representational content, since the emotion may be appropriate, or useful, even if the state of affairs represented does not obtain. But we can accommodate these emotions within our schema by taking them to be composed of two components: an affective component, which is non-representational, and a thought component, which has the content. The representational content of the emotion is given by the truth-conditions for the thought. In the instrumental sense of success, in contrast, the affective component will no doubt play a role in determining the usefulness of the emotional state.

One advantage in characterizing representational content in terms of truth-conditions is that it allows us to distinguish between two different aspects of a representation: its content and its *character*. If the content is whatever is represented, the character is the way in which that content is presented. In particular, the representation may be from a particular perspective or orientation in space, and this perspectival element is, for reasons we shall encounter later, best seen as part of the character of the representation—the way the content is presented—rather than as part of the content—whatever makes the representation true (e.g.).

There is, however, a kind of representation that does not fit easily into our schema: namely, what we might call 'bare' representation. Bare representations are those that represent an object without representing it as being a certain kind of way. So, for example, during a dinner-time conversation I may take a wineglass and use it to represent someone of our acquaintance, perhaps as part of an anecdote. ('He was standing *here*, and I was standing *there*, when suddenly, the door burst open and . . . .')

Positioning the wineglass on the table in relation to other objects may represent our mutual friend's actual location, but the wineglass itself just represents him, *simpliciter*, non-descriptively. Similarly, a proper name like 'Jeff' may represent an object, by referring to it, without describing it in any way. Some emotions too may be bare representations. I may simply fear the dark, or a particular wild animal, without fearing *that* anything. These bare representations are not characterizable in terms of truth-apt thought. In these cases the 'representational content' is simply the object represented, or the reference of the representation, not its truth-conditions.

Even if bare representation is not directly characterizable in terms of truth, however, it may be indirectly so. For bare representations are often (and, we might add, are always at least capable of being) components of representations *as*. As observed above, having stipulated that the wineglass is Jeff, I can move it around to represent Jeff as being in such-and-such a position. Proper names like 'Jeff' can similarly feature in assertions. The function of the bare representation in these cases is to make an object part of the truth-conditions of the larger representation (the representation *as*) that the bare representation is part of. So even bare representation is connected, though sometimes only counterfactually, with truth.

I have tried to characterize representation in terms of truth. What I have not done is provide a theory of representation. But we can perhaps begin to see what a theory of representation should do: it should explain *what it is* that connects the representational state with its truth-conditions. What is it that determines that this state should have *those* conditions (and hence *that* content)? It should also be a theory of character as well as content.

If this is what a theory of representation is supposed to do, then Donald Davidson's (1967) suggestion that meaning be defined in terms of truth-conditions is, if we identify 'meaning' in this context as representational content, less a controversial theory of representation than the application of an unexceptionable schema. It does not tell us (nor was it intended to) *how* sentences acquire the truth-conditions they do. Nor does it tell us anything about character. Paul Grice's (1957) account of meaning in terms of speaker's intentions takes us a little further down that road, since it aims to explain how certain linguistic expressions have the representational content they do—by being appropriately connected to certain mental representations. But it takes those mental representations for granted: it does not explain how they acquire *their* content. So neither Davidson nor Grice provides us with a theory of representation in the above sense, and, *a fortiori*, they do not provide rival theories of representation.

What determines their content is one question we can ask of representational states. Another, quite distinct, but not unrelated, question is what determines their epistemic status: why do some representational states count as states of knowledge, whereas others do not? Causal answers have been given to both these questions, and since causality is central to metaphysical inquiry, the more representation is explicable in terms of causation, the more likely it is that studying representation will tell us something about the nature of the world. So let us now take a look at the contribution that causation may make to our understanding of representation.

2.2 CAUSATION AND CONTENT

When, standing in front of the Eiffel Tower, you have a perceptual experience, what makes that experience an experience *of* the Eiffel Tower? Why, the fact that the Eiffel Tower caused your experience (by an appropriate route), surely. When you call to mind that unforgettable occasion in that charming Paris café four summers ago, what makes your memory a memory *of* that occasion? Why, the fact that the occasion in question caused a contemporaneous experience that in turn caused your memory—again, by an appropriate route (Martin and Deutscher 1966). Since perception and episodic memory so clearly involve our being causally affected by objects and events, indeed are paradigmatic instances of causation, it would be hard to take seriously a theory of perceptual and memorial representation that did *not* assign a central role to causation. Of course, as Paul Grice, one of the original proponents of the causal theory of perception conceded, more has to be said. Granted that perception and memory involve a causal tie, under what conditions does this causal tie determine content? Note the qualification 'by an appropriate route' above. Not just any old causal tie between a perceptual experience and the Eiffel Tower will make the Tower the object of the experience. Some causal chains between objects and experiences are 'deviant', sufficiently so to break the link between causation and content in these cases (Grice 1961). We will come back to this problem later.

As long as we focus on what was called in the previous section 'bare' representation, causation will be the obvious link connecting the representation and the item represented. But much mental representation is representation *as*: we do not merely perceive the Eiffel Tower, but we also perhaps perceive it *as* the Eiffel Tower (just think! the very thing we saw in the holiday brochure!), as very tall and as being of a dark colour. Now, if we take the view that content is wholly determined by the cause, then we will want to say that here we have, not just an object, but certain states of affairs causing the experience: the Tower's *being* the Eiffel Tower, and its being tall and dark causes my perception of it as the tall and dark Eiffel Tower. But this cannot be a necessary condition for my perceptual state's having the content it does, since the Eiffel Tower cannot fail to be the Eiffel Tower; yet one can see it without having the concept 'Eiffel Tower'. One might even have seen it as the Taj Mahal. And it could have failed to be tall and dark, yet one could still have seen it as tall and dark.

We have two issues here: one is the issue of conceptual content. Having an experience caused by an *F* does not guarantee that you experience it as an *F*, since you may lack the concept of *F*. The other issue is that of misrepresentation: you may see something as *F* when in fact it is not *F*. Can these commonplaces be accommodated by a causal theory of representation? Jerry Fodor (1987) suggests a way in which it might be done. Concept acquisition is a causal process, and acquiring the concept of, say, a horse involves contact with actual horses in appropriate conditions. And likewise for a large class of other concepts. As a first attempt, then, your idea counts as an idea of a horse by virtue of its being reliably caused by horses when in certain specified conditions. The difficulty that immediately arises is what Fodor calls the 'disjunction' problem. Perhaps my idea is reliably caused by both horses and cows, in which case my idea would count as an idea of 'horse or cow': a disjunctive idea. (Describing it as disjunctive is not meant to indicate anything about its internal structure: it need not be an idea made up of previously acquired ideas of horses and cows. The thought, rather, is that this is simply an idea that indifferently picks out both horses and cows.) But in some cases we want to say that you are not picking out a cow as a 'horse or cow' but, rather, misrepresenting a cow as a horse. That is, your idea is a non-disjunctive idea, just of a horse, and you are misapplying it to a cow. Fodor's ingenious proposal is to appeal to what he calls *asymmetric dependence*. When you misrepresent a cow as a horse, your token idea 'horse' is in this instance caused by a cow because ideas of that type are, in your case, typically and reliably caused by horses. In contrast, when a token idea 'horse' is caused by a horse, this does not depend on the fact that, in your case, ideas of that type are typically and reliably caused by cows. Where the idea is genuinely disjunctive, 'horse or cow', there is no such asymmetry of dependence. So we have here a causal account both of conceptual content and of misrepresentation.

So let us take a case where an object, say of next door's cat Rupert, causes in me a certain perceptual experience, the experience as of a tabby cat. My experience is of Rupert by virtue of being caused by him, and I experience Rupert as a tabby by virtue of possessing the tabby concept, which itself is reliably caused by actual tabbies, etc. But what if I misidentify Rupert? I may, because I see him from a distance in non-ideal conditions, mistake him for my own cat, Widmerpool, and thereby acquire the belief that Widmerpool is in the garden. Still, my experience is of Rupert, and my thought is of Widmerpool (and perhaps also of Rupert), all of which is explicable in causal terms. I am capable of having thoughts about Widmerpool by virtue of regular causal contact with him. On this occasion, my thought about Widmerpool is caused by seeing Rupert. It is not implausible to say that on this occasion I think (*de re*) of Rupert that he is Widmerpool, even though I do not think (*de dicto*) that Rupert is Widmerpool.

Now, sketchy though this account is, it looks fairly promising when we are dealing with the content of perceptual experiences and perceptual beliefs (beliefs acquired as a more or less direct result of perception). And since episodic memory begins with perception, it looks promising as an account of the content of episodic memories, too. The challenge for the causal theorist of representation in general is to extend the account to include such things as beliefs involving more theoretical concepts, ones further removed from ordinary perceptual experience, and singular thoughts about non-existent objects, including singular *negative* existential thoughts ('Gradgrind does not really exist'). Perhaps the causal theory of representation has some interesting consequences for our view of fictional characters and fictional truth. We might also ask, even though we have distinguished between the content and the character of thoughts (i.e. between what they represent and the perspectival element of the representation) whether the causal theory has anything to tell us about the character of thought. What is it, in causal terms, for a thought to be perspectival?

I beg leave to postpone these interesting and pertinent questions for another occasion (pausing only to refer again to Fodor (1987) for a discussion of the issue over theoretical terms), because my interest here is in the metaphysical significance of the causal theory of representation in precisely those cases where that theory is almost irresistible, namely perception and memory. When, in Chapters 8 and 9, we turn to fictional representation, the issues that arise do not (directly) involve causal elements of the representational mechanism.

One aspect of the causal theory of representation is that causality confers content, or at least is an essential component of what confers content. Can causality also confer epistemic status? To that question we now turn.

2.3 CAUSATION AND KNOWLEDGE: THE CAUSAL TRUTH-MAKER PRINCIPLE

There are circumstances, and happily they do not seem to be too uncommon, in which our acquisition of a belief about the world is so secure that we are content—at least, when we are not wrestling with sceptical objections—to count that belief as an item of knowledge. What circumstances are those, precisely? I am going to concentrate on the case of perceptual knowledge, as being arguably the simplest case. Clearly, the mere truth of a belief is not enough for it to count as knowledge: the fit between belief and world might be merely fortuitous. The traditional account of knowledge is in terms of *justified* true belief, but in the case of perception at least the requisite justification is not plausibly thought of in terms of reasons that could, at a moment's notice, be articulated by the knower. Winifred knows where a sound is coming from just by listening to it, and perhaps also by turning her head from side to side, even though she could not give an account of the reasons why this method works: she does not know, for instance, that it has anything to do with the minute differences in wave patterns of the sounds reaching her two ears. What matters is how the belief was acquired, whether or not the believer could give an account of it.

The causal theory of knowledge proposes that the acquisition of a true belief, if it is to count as knowledge, cannot be accidental. That is, the fact that the belief is true, given that it was caused in this way, is not purely coincidental. There is something about the way it was caused that makes its truth entirely unsurprising. What could that be? Surely, that somewhere along the causal chain leading to the belief is a state of affairs that also accounts for the truth of the belief (see Goldman 1967 for a classic statement of this position; another version of the causal theory is presented and defended by Swain (1979)). The causal chain, in other words, contains the *truth-maker* of the belief. Let us call this

The Causal Truth-Maker Principle: Perceptual beliefs that qualify for the title 'knowledge' are caused, in part, by their truth-makers.

The Causal Truth-Maker Principle (CTMP) is appealing, first because perception is evidently a causal process, and second because it successfully discriminates in very simple terms, and in a wide range of cases, between true beliefs that constitute knowledge and those that do not. For example, I see someone I take to be Julian, riding a trishaw down the street. I form the belief that Julian is riding a trishaw, and I happen to be right about that: Julian is indeed riding a trishaw at the very moment I form my belief, but the person I saw was his twin brother, Ivan. This a 'Gettier-type' counterexample to the thesis that knowledge is justified true belief. My belief is true, and it is justified, in this case by a perceptual experience in near-ideal conditions. So I have a justified true belief that Julian is riding a trishaw. But I do not *know* that he is. As Gettier (1963) pointed out, it is no good insisting that the justifying belief be itself justified, for this leads us into a regress. The CTMP, however, offers a way of avoiding regress. In this case, the truth-maker was not among the causes of my belief, which therefore fails to count as genuine knowledge. And so we can go on multiplying the examples.

Before we look at some difficulties for the CTMP, it is worth asking what, if anything, is the connection between a causal theory of content and a causal theory of knowledge. They do seem natural partners, and if one is impressed by the role of causality in one context, it seems reasonable to view it as having a role in the other. But there is no strict entailment either way. A causal account of content could be consistently combined with a non-causal account of what confers knowledge. And the CTMP does not entail a causal theory of content. Further, the causal relata required by the CTMP are rather different (in many cases at least) from those required by a causal theory of content. For instance, your belief that there are four trees on the hill in front of you may be caused by the state of there being four trees on the hill, and that may be why you *know* that there are four trees on the hill; but the kinds of causal connections that are supposed to confer the relevant content on your belief are rather more complex, involving previous encounters with hills and trees.

As we defined content in the previous chapter, however, there *is* a connection between content and the CTMP. For the content of a thought was defined in terms of its truth-conditions, and a truth-maker is what makes these conditions obtain on a particular occasion. Coincidence of cause and truth-maker is, then, what, according to the CTMP, confers knowledge.

It would, after all, be strange if there were *no* connection between determinant of content and determinant of epistemic status. Take a case where you acquire a very safe and non-committal perceptual belief about a particular object: *this object exists*. If anything is a case of knowledge, this is, and the combination of causal theories of content and knowledge explains why. The causal theory of perceptual content has it that your belief is about this very object by virtue of its being caused by that object. Since the conditions for perception are satisfied, the associated thought succeeds in being about that object. But this is all that it takes for the thought to be true. So the cause is also the truth-maker of the belief, and the belief therefore satisfies the CTMP's condition for knowledge. If perceptual knowledge was not a matter of the way in which the belief was caused, it would be puzzling why this particular belief was so secure. Of course, I have chosen a very simple example of the way in which cause confers content. Here the cause confers *de re* content: i.e. it makes the belief concern this particular object. It does not confer *de dicto* content: i.e. it does not explain why the belief is about a certain *kind* of thing. It would be rather harder to see the precise connection between the CTMP and the sophisticated account of *de dicto* content defended by Fodor.

For the purposes of the present project, it is the CTMP that is my main focus of interest.

How defensible is the CTMP? Counterexamples fall into two groups: those that call into doubt the *sufficiency* of the causal link with truth-makers to confer the status of knowledge on a belief, and those that call into doubt its *necessity*. It is not surprising if the CTMP does not represent a sufficient condition for perceptual knowledge. More likely, it will form the core of some more elaborate theory of perceptual knowledge. However, if it turns out that there are a number of problem cases that even an expanded causal theory of perceptual knowledge cannot deal with, then the CTMP loses some of its plausibility, even if the problem cases do not strictly conflict with its truth. So defenders of the CTMP would be well advised to show how the principle might be expanded to deal with threats to its sufficiency. Of course, if there are genuine counterexamples to its necessity as a condition for perceptual knowledge, then the principle is indeed in trouble.

Here, then, are the kinds of case that cause difficulties for the causal theory.

(a) First, are those against the sufficiency of the causal link with truth-makers.

(i) *Untrustworthy causes*. You are looking at a series of photographic slides that are being projected on to a screen. Unknown to you, some of the slides have been placed in the projector the wrong way around, so the screen image in these cases is left–right-inverted relative to the original subject. You are currently looking at a slide of, say, a street, and so form beliefs concerning the relative positions of the buildings, etc. Since this slide has not been inverted, your beliefs are true, and they are caused (ultimately) by their truth-makers, but they do not count as knowledge.

(ii) *The presence of relevant perceptual equivalents*. You are looking at a typical rural scene, and form the belief that there are sheep on the hillside. You are quite right about this. However, unknown to you, a farmer with a sense of humour has been placing cardboard models of sheep all over the hillside. As it happens, you really are looking at sheep, but you would not have been able to distinguish them from the models. Again, although the truth-makers and the causes of my beliefs coincide, those beliefs do not count as knowledge. (See Goldman 1976.)

(iii) *Deviant causal chains*. You believe that there is a dagger in front of you on the basis of a visual experience that is caused by a state of affairs precisely matching your belief, but via a 'deviant' causal chain. E.g. your experience is caused directly by stimulation of your brain by electrical impulses delivered by a machine operated by a scientist whose intention is artificially to replicate the kind of experiences which you would be having were you actually perceiving the dagger that is in fact in front of you. You believe, but do not *know*, that there is a dagger in front of you. (See e.g. Grice 1961; Strawson 1974; Peacocke 1979; and Lowe 1996 for detailed discussion.)

(b) Next are those against the necessity of the causal link with truth-makers.

(iv) *Beliefs with conventional conceptual content*. You see a tree, and, having some arborial expertise, form the (correct) belief that it is an Ilex oak. You can be said, in fact, to *know* that this is an Ilex oak. The truth-maker of the belief has to do with a conventional system of classification, however, whereas the causes of the belief are simply to do with the intrinsic properties of the tree.

(v) *Demonstrative thoughts.* You think, looking at the girl who has just walked into the café, 'That's Monica'. Since the person you are looking at *is* Monica, the truth-maker of your belief is that Monica = Monica. But this fact is a necessary one, and so cannot be the cause of your belief (on the grounds that causal relations entail counterfactual relations, and counterfactual relations can obtain only between contingently existing facts.) Your belief can nevertheless be classed as genuine knowledge.

A tempting way of dealing with the first case is to appeal to reliability. The process by which I form beliefs—looking at slides that may or may not be inverted—is simply not a reliable one. The example shows that whereas there are many cases where reliability can be cashed out simply in terms of a causal link with the truth-maker, there are others where this is not enough. Reference must be made to the tendency of the belief-generating process to produce true beliefs, or (to put it another way) to the high probability that the process will generate beliefs *via* their truth-makers. A similar proposal recommends itself in the second case. The presence of the model sheep on the hillside make it a matter of luck whether or not your belief is caused by a truth-maker. The belief-producing process may lead to true beliefs, as when you happen to be scanning a part of the hillside that has genuine sheep on it, but it does not do so reliably. Knowledge, in other words, is true belief acquired by a reliable method, which is one formulation of *reliabilism*. We will come back to the adequacy of this suggestion in a moment.

The third case introduces a new element, that of the deviant causal chain, which plagues causal theories of perception, memory, and action. In the case described, the deviancy of the chain undermines the experience's claim to be a genuine perception of the dagger. But is this a problem of knowledge? It may be possible to gain knowledge of the dagger by means of a visual or apparently visual experience, even if one cannot be said to be *perceiving* the dagger. And perhaps the criterion we use in deciding whether we have knowledge in this case will depend on the reliability of the process by which we come to have the belief, so that, once again, the case can be assimilated to the first. However, this may not go quite far enough. The fact remains that you mistake the immediate cause of your experience. You are not, as it turns out, really seeing the dagger. So how can you know that there is a dagger in front of you? Arguably, you cannot. But if this is so, then it suggests that reliability is not enough. The causal chain leading to a belief may be both reliable (in that the belief thus caused has a high probability of being true) yet deviant. So what distinguishes the deviant case from the non-deviant case in such a way that explains how non-deviant chains can lead to genuine knowledge? One proposal is to introduce the notion of *higher-order* reliability. A process may be reliable, as in the case of the honest scientist stimulating my brain, yet not itself have been generated by a process which reliably leads to true-belief-producing processes. Ordinary perception is not only reliable, it also exhibits higher-order reliability: our perceptual systems have evolved through a process of natural selection which itself makes it highly likely that perceptual systems will be reliable. Deviant cases, arguably, do not exhibit this higher-order reliability.

Or not always. We should note that a process can exhibit higher-order reliability and still be 'deviant' in the sense of being a non-standard route to perceptual knowledge. Deviancy *per se* is not a threat to perceptual knowledge. So, for instance, consider a device for the blind that converted light signals into electronic impulses that were channeled into the subject's brain in such a way as to produce an experience qualitatively identical to a normal visual experience. Provided that there was a reliable correlation between the nature of the experience and features of the subject's external environment, those experiences could count as providing genuine perceptual knowledge of that environment. Indeed, it would not be inappropriate to talk of the subject 'seeing' things.

Any attempt to address cases (i) – (iii) by appeal to the reliability of the belief-generating process does, however, have to address the following objection. In order to determine whether a belief was generated by a reliable process—one, that is, that made it likely that the belief was true—we need to fix upon the appropriate description of the process. But there is no one 'appropriate' description, since the process may be described at varying levels of abstraction, including, or excluding, certain details. Relative to some of these descriptions, the process may be reliable; relative to others, not. This is known as the 'generality problem' for reliabilism (see e.g. Conee and Feldman 1998). Consider again the sheep on the hillside. If we describe the process by which you acquired the belief that there were sheep on the hillside as simply 'looking at the hillside', then the process (in the circumstances) was not reliable. If, on the other hand, we describe the process as 'looking at a part of the hillside where there are no cardboard models', then arguably the process *is* reliable: in those circumstances, a perceptual belief that there are sheep on the hillside is likely to be true. Or consider the case of someone who is partially deaf listening to a conversation, but with the help of a fully functioning hearing aid. Is this a reliable way of acquiring beliefs about what is being said? Relative to the description 'listening when partially deaf', the process is not reliable, but 'listening when partially deaf but with a hearing aid that is sufficient to compensate for the partial deafness', it is reliable.

Would this relativity matter? If reliability is relative to a description, and we define knowledge as reliably acquired belief, then knowledge too is relative. Perhaps we could tolerate this, up to a point. 'Knowledge' is an honorific title. There is no clear cut-off point between cases of knowledge and cases of mere true belief. Where we draw the line will to some extent be a matter of convention. 'Knows that' is therefore unlike 'causes', which we suppose to be wholly objective. We have, by and large, an intuitive sense of the appropriate level of abstraction to apply in our description of a belief-acquiring process when it comes to deciding whether the belief counts as knowledge or not. But there will be some indeterminacy here, and that indeterminacy will—if the reliability account is correct—carry over into ascriptions of knowledge. But that knowledge ascriptions should exhibit some indeterminacy is exactly what we would expect.

But too much relativity would not be acceptable. Some ascriptions of knowledge are reasonable, others not, even if we can find some description of the process which makes it come out as reliable. We cannot simply ignore the generality problem. As Conee and Feldman point out, the problem arises because reliability is something that attaches to *types*. We have to ask, 'Is this type of process one that tends to lead to true beliefs?', which leaves us with the obligation to say how the relevant type is to be picked out. But perhaps we avoid talk of types altogether, and frame reliability in a different way. What follows is an alternative version of reliabilism that promises to do so, a view we might call the 'circumstantial dispositionalist account' of perceptual knowledge. According to this view, knowledge is the exercise of a disposition in the right circumstances (cf. Nozick 1981). In more detail, and including the CTMP:

x knows that p if and only if
(a) x has a true belief that p;
(b) x's belief is caused by the truth-maker for p;
(c) x is disposed to form the belief that p when p is true;
(d) x's belief was the manifestation of that disposition;

(e) the circumstances are such that, had *p* not been true, that disposition would not have been manifested.

According to this account, knowledge has both an internal and an external component. The internal component is the disposition to form the true belief that *p* in the presence of the truth-maker. A disposition is something that tends to persist, so a merely fortuitous acquisition of a true belief that *p* (perhaps as a result of a magician's intervention) is excluded from being a genuine case of knowledge. The external component is the circumstances in which that disposition is manifested. There must be a counterfactual dependence of the manifestation on the truth-maker for the belief. That counterfactual dependence is absent in case (ii). You are disposed to form beliefs about sheep in the presence of sheep, and in normal circumstances the manifestation of that disposition is counterfactually dependent on the actual presence of sheep. The disposition is not manifested in the presence of goats, for example. But in the peculiar circumstances of case (ii)—namely, the presence of cleverly designed cardboard models—that counterfactual dependence does not hold.

We have not characterized the process by which the belief is acquired except in very abstract terms. The only relevant question concerning that process is whether it involved a disposition to form true beliefs. It is presumably as objective and non-relative a matter as to whether someone is disposed to form true beliefs about, say, sheep, as it is whether something has the disposition to dissolve in water. We have not said anything about how that disposition is realized in terms of structural properties; nor need we. What about the circumstances in which the disposition is manifested? Should we not say something about these? If we had described the relevant circumstances as 'normal', we would certainly have invited the demand to say what counted as normal. But the circumstances in question are just those that do not threaten the counterfactual dependence of the disposition's manifestation on the truth-maker, whatever they are. And again, it seems to be just as objective and non-relative a matter whether the manifestation was counterfactually dependent on the truth-maker as it is that the explosion was counterfactually dependent on the short circuit. Perhaps there are difficulties with counterfactuals, but we face these anyway.

Now let us turn to the alleged counterexamples to the necessity of a causal link between truth-maker and belief. Consider the Ilex oak. Is this really a problem? *All* perceptual beliefs, let us concede, will have some conceptual content (assuming that they have a propositional structure; it may be that there are some perceptions lacking that structure that do not possess conceptual content—see e.g. Crane 1988—but that is not the issue here, since we are concerned only with perceptual *beliefs*). Those concepts will have their origins beyond the experience they help to structure. Does that matter? Given the admittedly conventional system of classification, all that is needed for your belief that this is an Ilex oak to be true are the intrinsic properties of the tree. And given that you possess the relevant concepts, and are suitably oriented vis-à-vis the tree, those same intrinsic properties will be sufficient to cause that belief.

In the case of the demonstrative thought, the obvious reply here is to deny that the truth-maker in question is Monica = Monica. There are, in any case, other grounds to question that Monica = Monica is the truth-maker, for the necessity of the truth-maker would appear to confer necessity on the belief it makes true; yet, 'That's Monica', being clearly empirical, does not look like a necessary truth. There are a number of candidates for the role of truth-maker for demonstrative thought, but one that would satisfy the contingency constraint relates perceiver and object: my thought is true because the person who stands in the appropriate perceptual relation to me is Monica.

The CTMP, perhaps suitably augmented, is, it seems, robust enough to deal with a number of apparent counterexamples (though later on in this inquiry we will see in what ways it needs to be qualified).

What we have arrived at is a sketch of mental representation, or at any rate of mental representation of a certain type, which makes use of two key notions: truth and causation. Both of these notions involve the world beyond the representation. But is there any reason to suppose that studying representations will tell us anything of philosophical interest about that world? The key idea here is that truth and causation combine somehow to transform beliefs into knowledge, and this is captured in the Causal Truth-Maker Principle, that perceptual beliefs qualifying for the label 'knowledge' are caused by their truth-makers. Although this is put forward as a piece of epistemology, it is also, I suggest, a significant metaphysical principle. We can ask of any perceptual belief, what, if it is true, makes it so? And what gave rise to it? Where the belief counts as knowledge, the CTMP requires that a sufficiently full answer to the second of these questions include the answer to the first. That is, the truth-makers of a large range of our perceptual beliefs have to be capable of being causally active. That may not seem terribly surprising or problematic, but the truth-makers of our perceptual beliefs about time (viz. tensed beliefs and beliefs about order and duration) are a matter of metaphysical debate. Some accounts of the truth-makers of those beliefs may not satisfy the CTMP, in which case we would have an interesting conflict between epistemology and metaphysics. If, on the other hand, we have a metaphysical account of those truth-makers (which is nothing less than a theory of time) that satisfies the CTMP, then, in so far as we are impressed by the CTMP, that is an argument in favour of the account. This line of argument will be prominent in Chapters 4 and 6.

First, however, I want to look at another feature of representation, one which is intimately connected with the metaphysics of time.

3

Egocentric and Objective Representation

> I fell into a highroad, for so I took it to be, though it served to the inhabitants only as a footpath through a field of barley.
>
> Jonathan Swift, *Gulliver's Travels*

3.1 PIAGET ON EGOCENTRISM IN THE CHILD

Representations of the world fall naturally into two kinds, each having a distinctive role to play in thought and action. The kinds in question are *egocentric* and *objective* representations. The first term marks representations that are sensitive to one's context or identity or, more narrowly, position in a given dimension (hence 'egocentric'), and the second marks those which are not sensitive in this way. For example, I can think of myself egocentrically as 'I', or objectively as the man in the white suit. The distinction between these two kinds of representation has a natural and unproblematic application to space, so we might expect the distinction to apply in a similarly straightforward way to time. When we attempt to apply the distinction to temporal representations, however, we encounter difficulties, difficulties which connect with a dispute about the metaphysics of time.

Here, then, are the questions I shall address in this chapter: What is it for a representation to be egocentric or objective? Can we make sense of the egocentric/objective distinction as applied to representations of time? This leads naturally to a question we shall address in the next chapter: Are there any reasons to suppose that an 'objective' representation of time actually plays a role in our ordinary thinking?

We will begin by looking at one very well-known application of the concepts of egocentricity and objectivity in developmental psychology.

For Jean Piaget, the shift from egocentric to objective thinking is characteristic of a certain stage in a child's cognitive development. What did he mean by this? We can illustrate his idea by a famous experiment. Children between the ages of 4 and 12 were presented with a pasteboard model of three mountains in a triangular configuration, each mountain readily distinguishable from the others in terms of shape, size, and colour. Having inspected the model from different perspectives, each child was shown ten pictures, each representing how the mountains would appear from different positions. The child was then stationed at a particular position facing the mountains. Finally, a doll was placed in some other position around the model, and the child invited to select the picture which represented how the mountains appeared *from the doll's point of view*. The result was that children of 8 or younger would typically choose the picture which represented how the mountains looked from the child's own point of view, *not* the doll's. Here is Piaget and Inhelder's comment:

[T]he children who invariably choose Picture I fail to realize that what they are expressing is actually their own viewpoint. 'That's the best one,' says Fer, 'he sees all three just as they really are!'. To see the mountains just as they really are—meaning, to see them as one sees them oneself. The phrase illustrates perfectly. . . . the unconscious character of the child's intellectual egocentrism. (Piaget and Inhelder 1956: 220)

Piaget's experiments have been much criticized, but I am concerned here less with the accuracy of his locating the shift away from egocentric thinking at around 8 years than with his conception of that shift. The egocentric child fails to recognize that other people, differently situated in space, will necessarily have different perspectives on what they perceive. The whole notion of a 'perspective', in other words, is not available to the child.

For Piaget, egocentricity is a general feature of the young child's conception of the world, not confined to one area of thought. So temporal thinking, too, can be egocentric. Again, an experiment will illustrate what Piaget means by this. The set-up consisted of two flasks: one was narrow and cylindrical; the other, placed immediately above the cylinder, had the shape of an inverted pear, its diameter at the widest point being larger than that of the lower flask. Water was poured into the upper flask (I), and could, by means of a tap, be made to run into the lower flask (II). Because of the difference in shape between the two vessels, a given drop in water level in the upper flask (I_1 to I_2) was accompanied by a rather larger rise in the lower flask (II_1 to II_2). The child was then asked to make judgements about the time it took for the water levels to rise or fall. Here are two representative cases (the child's responses appear in italics):

PEL (6 years): Does the water take as long to rise from here to there (II_1 to II_2) as it does to drop from here to there (I_1 to I_2)? *No.* How long does it take to rise here (II_1 to II_2)? *Two minutes, I think.* And to drop here (I_1 to I_2)? *Five minutes.* Why? *It's bigger and there's more water on top.*

CHAP (7 years, 4 months). The water is allowed to drop from I_1 to I_2: Did you see that? *Yes, the water dropped here* (I_1 to I_2) *and rose there* (II_1 to II_2). Did it take the same time here (I_1 to I_2) and there (II_1 to II_2)? *No, it took longer to rise in here* (II) *than to drop down there* (I). Why should it take longer in here (II)? *Because I could tell.*

(Piaget 1969: 36–7)

For these children, suggests Piaget, enlarging on their own rather laconic explanations, time is *plastic*: it expands when the movement of water is slow, and contracts when the movement is fast. A greater displacement of water, or, more generally, a greater amount of activity, must, on this view of the world, take more time. The following conversation with the same 7-year-old provides another illustration of the point:

How long does it take you to go home? *One hour.* And if you are in a hurry? *I go more quickly.* Does that take more or less time? *More time.* Why? *Because.* (Piaget 1969: 38)

In describing these responses as egocentric, Piaget clearly takes them to be analogous to the phenomenon revealed in the mountains task. At some level, of course, the two kinds of result have something in common: they suggest a general failure on the child's part to distinguish appearances from reality. But perhaps more striking than the similarity is the *dis*similarity between the two cases. The alleged failure in the mountains task is the failure to understand that our view of the spatial relations between things is necessarily perspectival. In the water task, by contrast, the child is not being asked, and failing, to co-ordinate different perspectives on time—at least, not in a way that is analogous to the spatial case. Rather, the failure, or apparent failure, involves an incorrect understanding of the relationships between motion (or change in general), speed, and duration. We can represent this by contrasting the egocentric inference the child seems to be making concerning speed and duration:

x was faster than y
therefore: x took more time than y

with the correct inference:

x was faster than y
therefore: x took *less* time than y (to reach the same point)

If this is the right interpretation of the child's thought, then the egocentric inference above will disrupt the child's understanding of another relationship: that between simultaneity and duration. Here is the correct inference:

1. x started at the same time as y
2. x finished at the same time as y
therefore: x took as long as y

Piaget found that there were certain kinds of tasks where children of around 6 or 7 readily assented to the premises (and appeared to understand them), yet failed to draw the correct conclusion. These tasks included those in which x and y proceeded at different rates.

Again, I am neither endorsing nor rejecting Piaget's view of the child's cognitive development. My purpose in discussing him is to distinguish a specific kind of egocentricism in representations of the world within the more general phenomenon which Piaget believes he has identified. There are two distinct senses in which the child's representation of, or thinking about, space is 'egocentric', as Piaget uses the term:

(i) the representation is sensitive to the child's location in space;
(ii) the child is not aware that it is sensitive in this way; that is, she or he does not grasp the distinction between location-sensitive and location-insensitive representations.

Whatever one may think of Piaget's experimental procedure, this much is clear: the child's representation of the spatial position of the mountains is egocentric in the specific sense that it is sensitive to his or her location in space. And we can contrast this with a representation of the mountains which is not egocentric in that sense, and which the co-ordination of different perspectives perhaps requires. We can agree with Piaget, then, that a full conception of space must make room for this distinction between location-sensitive and location-insensitive representations, whether or not we agree that children of a certain age do not grasp this distinction.

What concerns me is whether the representations of time that both children and adults use in everyday life are egocentric in sense (i); whether there is a representation which is genuinely temporal, yet not egocentric; and whether a full conception of time requires the kind of distinction made in (ii). So the idea of objective time which I am pursuing is not quite Piaget's. His conception of objective time is expressed in the following passage:

[O]bjective time is homogeneous and uniform, and can only be grasped by the decentration of thought away from subjective ideas of duration. Hence, in our experiment, the child cannot simply project his own time into each of the motions in turn—a characteristic of egocentric intuition during the first two stages—but must conceive of a homogeneous time common to two motions, i.e. independent of the velocities of either. (Piaget 1969: 48)

Part of this is the idea of a single time series, along which different activities are located, and by means of which the duration or velocity of those activities can be meaningfully compared. This idea, the temporal unity of events, is one to which we shall return in Chapter 9.

3.2 EGOCENTRIC AND OBJECTIVE SPACE

Before proceeding any further, however, we need a more refined account of what it is to be an egocentric representation of space, since it is the temporal analogy—or lack of it—with space, that is my concern. The basic idea is captured by Gareth Evans:

The subject conceives himself to be in the centre of a space (at its point of origin), with its co-ordinates given by the concepts 'up' and 'down', 'left' and 'right', and 'in front' and 'behind'. We may call this 'egocentric space', and we may call thinking about spatial positions in this framework centring on the subject's body 'thinking egocentrically about space'. (Evans 1982: 153–4)

But, as Evans realizes, an egocentric (representation of) space is not one which just happens to be centred on one's own body. As John Campbell puts it:

[N]ot just any way of thinking about the subject will do. The notion of egocentric space is a psychological notion; the reason we want it is to explain why the infant, for example, turns one way rather than another. In particular, perceptual knowledge of the body will not do. Merely seeing one's own body in a mirror, for example, and using it to set up a system of axes will not provide one with an egocentric frame. (Campbell 1994: 10)

What else is needed? There is an intimate connection between a genuinely egocentric space, on the one hand, and action, on the other. As we move about the world, perceptual information guides our movement in a direct way. Seeing the pen over there, I know, without having to go through any conscious calculation, or acquire any further information, just how to move my hand in order to get it. Egocentric information is 'immediately' action-guiding ('immediate', that is, in the sense of not issuing in action via some further conscious interpretation of the data; there will be any number of causal intermediaries), and given this immediacy, there is no reason to deny that babies and non-human organisms, for instance, have egocentric representations of space. So, to summarize: an egocentric representation of space is one that

(a) is essentially subject-centred—i.e. the spatial locations of objects are defined in terms of their position vis-à-vis the subject's body;
(b) provides behaviourally relevant information of an immediate kind—e.g. it would dispose a subject who desires a certain object in its visual field to move in a certain way in order to obtain that object.

A consequence of (a) is that egocentric spatial locations of objects will vary as the subject moves around in space, where as (b) articulates how such a representation would be manifested. It is not a requirement of egocentric representation that the subject possess the concepts of space or the self, even though we would naturally present the content of such representation in terms of those concepts: all that is required is a certain sensitivity to the environment.

We might at this stage note two further quite contingent features of the egocentric representation of space:

(c) the egocentric information provided by perception is not confined to one position: objects in the visual/auditory field can be here, there, to the left, etc.;
(d) the subject's behaviour is similarly not confined to one egocentric position.

Both of these features are likely to be significant in explaining how we come to grasp the distinction between an egocentric and an objective representation of space. Having defined the former, we can define the latter negatively. An objective representation of space, then, would be one that

(a*) need not, though may be, centred on the subject's body (and hence objectively represented spatial locations need not, though may, vary as the subject moves through space);

(b*) does not provide behaviourally relevant information of an *immediate* kind, though it may become relevant with further information (see Perry 1979).

Do we ordinarily have access to such a representation? I think it is pretty clear that we do. Paradigm examples of objective representations of space are not hard to find: a street map, an anatomical drawing, a model of a DNA molecule—all these represent the relationships between objects (buildings, streets, organs, atoms) in a way which is insensitive to one's own position. They are, or are very nearly, perspectiveless. (I say 'very nearly' because both the street map and the anatomical drawing are two-dimensional projections, and so in a sense imply a viewpoint.) More controversially, we could propose that objective representations are internalized as 'cognitive maps': we find our way around town by means of a mental representation which is spatially objective. We cannot simply translate objective information into egocentric information, unless somewhere in this representation we locate ourselves—a mental 'you are here' sign. ('Hobb's Lane is at right angles to the High Street and to the south-west of the *Three Tuns*. I'm standing outside the pub now, so I need to turn left, then left again.') Apart from our ability to navigate our way around places, we may cite as evidence for an objective cognitive map the fact that we engage in reciprocal communication with subjects who have different spatial perspectives and so do not share our egocentric space. We might, for example, be giving instructions to someone moving about on the sea-bed, or in space, on the basis of what we can observe of him and his environment on a screen. Here we need to reconcile different egocentric representations, and since we cannot simultaneously entertain two such conflicting representations, the best explanation of our ability to communicate with others in these situations is that we map their positions on to an objective space.

What of temporal representation?

3.3 IS THERE AN EGOCENTRIC/OBJECTIVE DISTINCTION FOR TIME?

Initially, it seems that nothing could be easier than constructing the distinction between egocentric and objective temporal representations, since we can exploit familiar analogies with space. However, when we try to explore the analogies further, a number of difficulties emerge, difficulties which threaten to undermine the very idea of an egocentric/objective distinction for time. The difficulties concern, respectively, the metaphysics of time, what it means to treat a temporal representation as perspectival, and the possibility of a tenseless language. I shall deal with these in turn. First, however, the analogies.

3.3.1 The analogies

These are best illustrated by means of the terms we would use to express our mental representations (see Table 3.1). It is assumed here that the variables on the right-hand side of the table would not be replaced by egocentric expressions, such as 'me', 'my line of vision', or 'the present moment', for 'adjacent to me', 'at right angles to my line of vision', and 'simultaneous with the present moment' would be egocentric representations (allowing for the moment that we can talk of temporally egocentric representations).

Table 3.1: Egocentric and objective expressions

	egocentric	*objective*
space	here, there to the right down	adjacent to x at right angles to x four miles from x
time	now, then past, present, future yesterday, next year	before y simultaneously with y at 1500 hrs GMT on 9/2/13

The analogies between the upper and lower parts of the table are clear enough. *Tensed* terms such 'now', 'yesterday', etc. reflect the subject's changing position in time, just as spatially egocentric terms ('here') reflect the subject's changing position in space. And information of the form that such-and-such is happening now is immediately relevant to behaviour in the same way as information that such-and-such is happening here. Indeed, it seems that we cannot have spatially egocentric information that is not also temporally egocentric. But, crucially, there are some disanalogies too. The information provided by perception appears to be confined to one temporal position: events in the visual/auditory field can only be experienced as occurring *now*. The subject's behaviour, or at least the behaviour that is under immediate temporal control, is similarly confined to one temporal position.

Further disanalogies emerge when we consider the case outlined in the previous section for an objective cognitive map. *Reciprocal* communication is not (normally) possible between subjects who have different temporal perspectives—we share a common now. So there is typically no need, in such communication, to reconcile different egocentric times. This is not to say, however, that we are not aware of different temporal perspectives. Memory, for instance, provides us with awareness of different egocentric times. Think of your first day at school, for example. Though now past, it can also be present, before your mind's ear and eye (and maybe nose too).

Whatever the disanalogies between spatial and temporal representations, there are clear differences between the temporal terms on the left-hand side of the table and those on the right, just as there are clear differences between the corresponding spatial terms. One way of characterizing the difference is that those on the left are *tensed*: in using them to describe an event, we locate it in the past, present, or future. Those on the right, in contrast, are *tenseless*: in using them to describe an event, we do *not* thereby locate it in the past, present, or future. And, surely, we do use tenseless representations of time whenever we use calendars and diaries. Are these not objective representations, just as much as an A–Z map of London? I do not, as time passes, need constantly to alter the dates of my engagements in my diary (not if I am organized, that is). And if the date of our meeting was 26 October 1998, then 26 October 1998 it will have remained, despite our subsequent inexorable journey down the river of time. Or take the following assertion: 'The date of the launch is five days after the conference.' This certainly does not appear to have any implications for our position in time vis-à-vis the events in question. Nor does it provide action-guiding information of an immediate kind. If I wish to be at the launch, for example, the assertion does not on its own tell me whether I need to make preparations now, or that I am too late to do so. So, in terms of our characterization of objective representations, the assertion would appear to be an objective representation of the time of the events.

Even conceding that there are significantly different kinds of temporal expression, however, it does not follow that those on the right are objective in the way that their spatial counterparts are. Consider the following remarks. According to J. R. Lucas, '[i]t is part of the concept of time that it is connected to us, whereas it is not absolutely necessary . . . that space should be connected to us. The essential egocentricity of time is reflected in the ineliminability of tenses' (Lucas 1973: 280). According to Arthur Prior 'it is only by tensed statements that we can give the cash-value of assertions which purport to be about "time" ' (Prior 1967: 198–9). And Michael Dummett contends that 'what is temporal cannot be completely described without the use of token-reflexive expressions' (Dummett 1960: 356). What is going on here? And how do we square these assertions with diaries, calendars, and other apparently tenseless means of fixing the times of events?

The thought on which these remarks converge is that there is something fundamental about the egocentric representation of time: that terms like 'now', 'then', 'yesterday', 'next year', and so on, are reflective not merely of our contingent temporal perspective on events, but rather of some intrinsic feature of time itself. So any putatively 'objective' representation of time, that included no tensed expression, would either not be a representation of time at all or an incomplete one. Here a thesis about temporal representation is bound up with a metaphysical thesis about time itself. Let us examine the metaphysical thesis in more detail.

3.3.2 The A-theory and the B-theory

The very idea that we can extend the egocentric/objective distinction to time is tied up with a particular view of the metaphysics of time. That, in large part, is why it is so controversial. The view in question concerns the reality, or otherwise, of the passage of time. On this issue there are, broadly, two theories. The *A-theory* holds that time does in reality pass, and that events are located in what McTaggart (1908, 1927) called the 'A-series': that is, they are located in the more or less distant past, present, or more or less proximate future. These positions are subject to constant change, so that what was once present becomes increasingly past. That an event is e.g. present is an entirely mind-independent matter, requiring no one to entertain or give utterance to the belief that it is present. (A number of philosophers who would describe themselves as A-theorists would take exception to at least part of my characterization of their position, but this is a good enough starting point for discussion; variations on the A-theory will be discussed later in this chapter and in the next.) The *B-theory*, in contrast, denies that time passes in this sense. Nothing in reality changes its position in time. Rather, events are located in what McTaggart called the 'B-series':

that is, they are earlier than, simultaneous with, or later than other events or times, and these relations between particular events remain absolutely constant through time. We may truly represent an event as present, but what makes this representation true is not its location in the A-series, the mind-independent presentness of the event, but rather the unchanging B-series relation between the time of the representation and the event represented. So, for instance, my stated judgement that it is now snowing is true if and only if the event of snowing is simultaneous with the time I make my judgement. There is no A-series in reality; there are merely representations of A-series positions. (See Smith 1993 for an extensive articulation of both theories, in their various guises, and a thorough and sophisticated defence of a version of the A-theory; for defences of the B-theory see Oaklander 1984 and Mellor 1998.)

There is a structural asymmetry between the A-theory and the B-theory in that, although the B-theory rejects the reality of the A-series, the A-theory does not (standardly) reject the reality of the B-series, or relegate it to a feature of mental representations. For the A-theorist can plausibly argue that the B-series supervenes on the A-series. The following are some of the A-theoretic analyses that have been offered:

x is earlier than y if and only if:
(1) whenever y is present, x is past (cf. McTaggart 1927: 271)
(2) (x is present and y is present) or (x is present and y is future) or (x is past and y is future) or (x is more past than y) or (x is less future than y) (Gale 1968: 92)
(3) it either is, was, or will be the case that both x is past and y is present (Geach 1979: 98).

And we might add the following:

(4) x is n units past and y is m units past, and $n > m$, or x is v units future and y is w units future, and $v < w$.

Each of these embodies the idea that the B-series supervenes upon the A-series. But the B-theorist, in contrast, is not offering a supervenience account, but rather *rejecting the reality* of the A-series. The immediate challenge for the B-theorist, then, is to provide a semantics for temporal statements. In particular, the A-theory needs to provide a semantics for statements about A-series position. Here is one attempt:

Date B-theory semantics:
A token of 'e is present', tokened at B-time t, is true if and only if e occurs at t.

A token of 'e is past', tokened at B-time t, is true if and only if e occurs earlier than t.

A token of 'e is future', tokened at B-time t, is true if and only if e occurs later than t.

A 'B-time' is simply a position in the B-series, which, unlike A-series positions, does not shift, and will typically be expressed by a date.

In one sense, the B-theory regards the A-series as mind-dependent. But it is important not to interpret this as the mind-dependence thesis Richard Gale attacks in *The Language of Time* (1968). As he characterizes it:

> The becoming of physical event P_1—its being future, then becoming present and then past—is analysed in terms of the fact that it is later than mental event M, simultaneous with a subsequent perception of it, namely M_1, and earlier than later mental event M_2. (Gale 1968: 231)

But what of the mental events themselves? In what sense are they present? Gale continues:

> It cannot be claimed, without vicious circularity, that a mental event, such as an act of perceptual awareness, is present only in the elliptical sense of being related to some physical event of which it is a perception; for it has already been contended that a physical event is only present in the elliptical sense of being the object of some perception. Nor can it be claimed, without instituting a vicious infinite regress, that a mental event is present only in the elliptical sense of being related to some mental event in a second-order series of mental events.

The conclusion is that mental events must be past, present, and future absolutely: i.e. that they form a true A-series. This mind-dependence account is an unattractive amalgam of the A-theory and the B-theory. And, as Gale points out, it leads, among other things, to a problem of mental/physical interaction. Physical events appear to cause mental ones. But if physical events lack intrinsic presentness, how can they cause a mental event to become present?

Fortunately, there is no need for the B-theorist to embrace this strange mind-dependence view. We should distinguish between *representing* A-series position and *instantiating* it. Talk of being present 'in an elliptical sense' obscures this crucial distinction. On the B-theory, nothing instantiates A-series position, though beliefs can represent it. Events, whether mental or physical, can be correctly described as being present, and the correctness of such descriptions depends only on B-series facts.

The B-theory semantics mirrors the semantics we would offer for spatially indexical expressions (and no one is troubled by the 'mind-dependence' of hereness). For example:

A token of 'o is here', tokened at place p, is true if and only if o is located at p.

A semantics for spatial expressions that mirrored the A-theory semantics would, in contrast, look needlessly complex. Consider, for instance:

x is between y and z if and only if (x is to the right of y and to the left of z) or (x is to the left of y and to the right of z) or (x is behind y and in front of z) or (x is behind z and in front of y) or (x is above y and below z) or (x is above z and below y).

What is wrong with this analysis, which otherwise correctly represents the correlation between betweenness and the other spatial relations, is that it analyses a non-perspectival relation (between) in terms of perspectival ones (to the left of, behind, above). The latter depend on our position, or the position of whoever is ascribing those relations to groups of objects. The former is not so dependent.

Why can we not raise a similar objection to A-theory semantics? Why can we not say that it is inappropriate to analyse a non-perspectival relation (earlier than) in terms of perspectival properties (past, present, and future)? The objection is unlikely to impress proponents of A-theoretic semantics, who will say that 'past', 'present', etc. are quite unlike 'to the left of' and similar terms. But how exactly will they articulate this? What one might expect is a denial that tensed terms are perspectival; for if the A-theory is correct, then a tensed description of reality might reasonably claim to be perfectly 'objective', in the sense of describing how things actually are, rather than merely how they appear from a particular perspective. If I have a changing perspective on time, it is because I am being carried along by the passage of time. To say that the truth of tensed judgements depends on my perspective is to locate the dependence in the wrong place. The remarks by Lucas and Dummett quoted above, however, imply that tenses *do* reflect our position. Moreover, they are both perspectival (in some sense) *and* objective. But how is it possible for one and the same aspect of a representation to both reflect our perspective on time and capture an intrinsic feature of time, one that does not depend on our perspective? There is at least the appearance of paradox in this. To see if there really is a contradiction here, we need to look more carefully at the notion of perspective.

3.3.3 Content, truth, and context

What is it for a representation to be 'perspectival'? Adrian Moore has offered this helpful characterization (although he goes on to refine it in the light of a possible counterexample): 'a representation whose type does not determine its content' (Moore 1997: 11). Perhaps we should amend this to 'a representation whose type does not wholly determine its content', since the intrinsic character of the representation—revealed by the words we would use to express it, in the case of a purely mental item—is clearly going to be relevant to the determination of its content. So what else is needed? The context in which the representation is entertained or put forward. If, following the suggestion made in Chapter 2, 'content' can be replaced by 'truth-conditions', we can propose this definition of a perspectival representation: a representation whose truth-conditions include the context in which the representation is entertained or produced, *by virtue of the fact that it is the context of the representation.*

Egocentric spatial representations, like 'it's hot here', are clearly perspectival in this sense. The context of the utterance is the place in which it is made, and the representation is true just in case it is hot at that place. Non-perspectival spatial representations can be true of their context, as for instance when I say that the Moon is *n* miles away from the Earth when I am actually standing on the Moon; but the fact that this is the context of the utterance is incidental to its truth: my utterance would have been true even if I had been standing somewhere else.

Now, suppose that the A-theory is correct. Does it follow that representations of the form 'It's snowing now' are not temporally perspectival? Not necessarily. It depends on the account we give of the truth-conditions.

According to what we may call the 'simple' A-theory, the truth-conditions of tensed representations are as follows:

Simple A-theory semantics:
A token of 'e is present/past/future' is true if and only if e is present/past/future.

Here the truth-conditions are quite independent of the truth of the token. In fact, there is no implication that the token is located in time at all. Of course, any such token will in fact be located in time, and any present-tensed token will concern its temporal context; but its truth does not, on this account, depend on the fact that this *is* its context.

Contrast this with a more complex A-theory semantics, one that pays attention to the time at which the token was tokened:

Date A-theory semantics:
A token of 'e is present', tokened at A-time t, is true if and only if e occurs at t.
A token of 'e is past', tokened at A-time t, is true if and only if e occurs earlier than t.
A token of 'e is future', tokened at A-time t, is true if and only if e occurs later than t.

Thus, for instance, a token of 'e is present', tokened 24 hours ago, is true if and only if e occurred 24 hours ago. These schemata are, of course, structurally identical to those offered by the B-theorist. The crucial difference is that in A-theory semantics the variable t ranges over A-times, that is, positions in the A-series. In this more complex semantics, the temporal location of a token enters into the token's truth-conditions; moreover, it is by virtue of being the token's location that it does so.

The advantage of the date semantics over the simple semantics is that it delivers a more intuitive result when we come to consider the truth-value of past tokens. Suppose on Monday I say, truly, that it is snowing. On Tuesday, when it is no longer snowing, I want to assess the truth-value of my earlier utterance. The simple semantics says that Monday's token is now (on Tuesday) false, since it is not snowing. The date semantics says that it is still true, because it was snowing at the time I uttered it, and this is surely the more sensible thing to say. On the date semantics, but not the simple semantics, the truth of tensed representations is sensitive to their position in time, and so they count as 'egocentric'. On the simple semantics, the content of a tensed token does not change with its position in the A-series, though its truth-value does. On the date semantics, content does vary with A-series position, but the truth-value remains constant.

What happens, though, when we put the date semantics together with the supervenience thesis embodied in analyses (1)–(4) above? It looks, on the face of it, as if circularity threatens. The threat is only superficial, however. We can substitute any of the analyses for the B-series terms on the right-hand side of the date semantics. Here are some examples:

A token of 'e is past', tokened at A-time t, is true if and only if one of the following disjuncts obtains: e is past and t present, e is more past than t, etc.

A token of 'e is past', tokened at A-time t, is true if and only if the following conditional obtains: whenever t present, e is past.

The upshot of this exploration of different A-theoretic schemata is this. If we accept the simple A-theory semantics, then tokens of 'e is past' and the like are *not* perspectival, in the sense defined above. If the simple semantics were all that was available to the A-theorist, there would then be a conflict between saying that tensed representations both capture an intrinsic feature of time itself *and* are perspectival. If, however, the A-theorist adopts the date semantics, then there would be no such conflict.

Can we, then, distinguish between egocentric and objective representations of time? It is natural to identify 'egocentric' with 'perspectival'. In which case, we should say that the egocentric/objective distinction with respect to time is available for the 'simple A- theorist', but not for the 'date A-theorist'. But even within the simple semantics, we can distinguish between those temporal expressions which intimate one's position in time vis-à-vis the events spoken of and those that do not. For it is part of the A-theorist's way of looking at the world that we are (while we are alive at least!) necessarily located in the present. Arguably, this is something that the A-theorist is obliged to account for (see Bourne 2006: ch. 1 and Braddon-Mitchell 2004). Assuming that they can account for it, the A-theorist can appeal to this fact in explaining why 'e is past' intimates the fact that one is speaking from a time that is later than the event in question. Given a simple semantics, the statement by itself does not entail anything about the temporal location of the speaker: it does so only in conjunction with the inescapable (though not analytic) truth that the speaker is in the present. It is in these terms, I suggest, that we should understand Lucas's remark that '[t]he essential egocentricity of time is reflected in the ineliminability of tenses' (Lucas 1973: 280). 'Egocentric' temporal representations are instanced by ordinary tensed assertions, such as 'It was snowing yesterday', whose content reflects an objective, mind-independent A-series, and which intimate the assertor's position vis-à-vis the events spoken of. 'Objective' temporal representations are instanced by assertions containing only B-series expressions. Reasons to think that we can draw the egocentric/objective distinction for temporal representation do not, therefore, automatically favour the B-theory. Nevertheless, the A-theorist who allows the distinction will hold that the egocentric representations are primary, and the objective ones derivative.

To summarize the story so far, the difference between the A-theorist and the B-theorist consists in their distinctive view of three issues: (i) an ontological issue concerning whether or not the B-series supervenes on the A-series (in a full-blooded sense of supervenience, according to which the subvenient properties ground or determine the supervenient ones); (ii) a corresponding epistemic issue concerning whether or not 'objective' representations of time are dependent upon more fundamental egocentric representations; (iii) a semantic issue concerning the relationship between tensed beliefs/assertions and the world. It might be thought that a study of temporal representation can tell us little about the ontological dispute between the A-theory and the B-theory, but in fact issue (ii) turns out to be crucial. For it is essential to B-theorists' articulation of their position that B-series expressions and concepts are not derived from A-series ones. There are, as we might alternatively put it, genuinely tenseless temporal representations. But some A-theorists (or at any rate those hostile to the B-theory) have asserted that a tenseless language or way of thinking is impossible. To that issue we now turn.

3.3.4 The possibility of a tenseless language

Few would doubt that there are genuinely tenseless sentences, of which mathematical and logical assertions are examples. There is simply no point in construing 'The square root of 49 is 7' or '$p \rightarrow (p \vee q)$' as having a time-sensitive truth-value, or as locating their subjects in the past, present, or future. (Unless, perhaps, we take a certain kind of intuitionist view of mathematical statements, on which such statements are true only when there is a known proof.) The contentious issue is whether tenseless statements can really be *temporal*.

Now we are entering this controversy, we should be more careful in our conception of what it would be for a temporal assertion to be genuinely tenseless. An assertion is tenseless in the minimal sense if it does not imply that the state of affairs of which it speaks is in the present, as opposed to the past or future (or in the past, as opposed to the present or future, etc.). And here there seems little reason to deny that there can be tenseless temporal assertions in this minimal sense, for such assertions can be formed from disjunctions of tensed statements:

Either Eric witnessed the incident *or* he is now witnessing the incident *or* he will witness the incident.

This is clearly neutral on the question where in the past, present, or future Eric's witnessing the incident is to be located. However, it is *not* neutral on the question whether or not Eric's witnessing the incident is to be located in one of the past, present, or future. A fully tenseless assertion would not imply even this, that Eric's witnessing the incident had some location in the past, present, and future. And the possibility of tenseless assertion in this more than minimal sense is just what the B-theorist needs to articulate their position, for the B-theorist denies that anything is objectively located in the past, present, or future. So can there be tenseless assertions in this sense?

Consider again the assertion 'The date of the launch is five days after the conference'. In the mathematical context we might be happy to construe the copula as tenseless, but is it so here? The natural assumption to make, on hearing this remark, is that the launch lies in the future. The thought expressed is thus a tensed one. If the launch now lay in the past, the assertion would at least be rather misleading; the speaker should have said 'the date of the launch *was* five days after the conference'.

The point here may go rather deeper than a simple gesture to the conventions of English grammar. Roger Teichmann (1998) has suggested that it would be impossible to construct a genuinely tenseless temporal language, on the grounds that grasp of such a putatively tenseless language would necessarily be parasitic on grasp of tensed expressions. Consider a candidate tenseless expression:

It rains on 4 August 2005.

This appears to be a structured sentence from which we may abstract the first part to form an assertion in its own right:

It rains.

This gives no clue as to when in the past, present, or future, raining takes place. However, unlike a timeless expression like 'The square root of 49 is 7', it implies that the state of affairs in question is located somewhere in time. What would constitute grasp of such an expression? Here is Teichmann's answer, and the key element in his argument against a tenseless language:

To grasp the meaning of the tenseless 'It rains' *in the way in which we grasp the present-tensed* 'It is raining', a person would have to be able to distinguish the truth of that sentence from its falsehood. But how could someone ever distinguish the case where it (tenselessly) rains from that where it does not? The truth at a place of the tenseless sentence 'It rains' consists in its having once rained there, or its going at some time to rain there, or its raining there at the moment. A person could of course see that it was at that moment raining, and might on that account say 'It rains'. But how could he ever manifest an ability to recognize that it is not (tenselessly) raining? It would be impossible ever to recognize the falsehood of the tenseless 'It rains', for it would be impossible to know that it had ever rained there, or that it would never rain there, in part because of the unrestricted form of this proposition ('never'), and in part because of the impossibility of knowing of the future. (Teichmann 1998: 182)

To which the following responses could be made:

First, the general principle being appealed to here is very close to, if not identical with, the verificationist dictum according to which an assertion is meaningful only in so far as it can be verified or confirmed. This does not by any means defeat the principle, but it does highlight its controversial nature. The italicized phrase in the above passage, however, may indicate that the principle is not intended to apply in all contexts.

Second, the very same line of reasoning would rule out as meaningful 'It is raining somewhere in the universe', since it would be impossible ever to establish that this assertion was false. This assertion, like 'It rains', is unrestricted, though the impossibility of knowing the future (in so far as this is a result of the ontological indeterminacy of the future) has no spatial counterpart. But if it is the impossibility of knowing an indeterminate future that is the real problem, then the indisputably tensed 'It will be raining' is equally problematic.

Third, the argument would also rule out some of the A-theoretic analyses listed in 3.3.2. Take McTaggart's analysis, for instance. Under what circumstances would we be able to judge as false 'whenever x is present, y is past'? We are limited, in fact, to the disjuncts of Gale's analysis (and, as just noted, any of them involving future tenses will be problematic).

Fourth, suppose that 'x is earlier than y' were to be understood as parasitic on tensed assertions. Then it would be quite unclear why we would judge Gale's analysis 'either x is past and y is present or x is present and y is future . . . etc.' as anything other than an arbitrary disjunction. Consider this question: On what basis do we judge the truth at one time of one of the disjuncts to entail the truth at other times of the others? What grounds the thought that each disjunct has something in common with every other? The obvious answer is that each disjunct preserves the *B-series order* of the events in question, and this answer is clearly informative. If our understanding of 'x is earlier than y' were parasitic on our grasp of the tensed disjunction, however, this answer would be completely uninformative.

3.4 CONCLUSION

In this chapter we have introduced and explored in detail two kinds of representation, which we labelled 'egocentric' and 'objective'. As generally understood, an egocentric representation is one that reflects a perspective on reality, more specifically one that is, literally, self-centred. An objective representation, in contrast, is one that attempts to represent reality as it is in itself, from no perspective. In the case of space we have, it seems, no difficulty in making this distinction. In the case of time, matters are less straightforward. It has been suggested, by John Lucas for example, that we cannot think of time except in egocentric terms. But if 'egocentric' is being contrasted with 'objective' here, this suggests, puzzlingly, that we cannot think of time as it is in itself, but only in perspectival terms. Part of the purpose of this chapter, then, was to dispel the air of paradox around the idea of an aspect of our temporal representations that both represents an intrinsic feature of time and reflects our temporal perspective. This has meant clarifying the terms 'egocentric', 'objective', 'perspectival', and 'tenseless'.

Exploring the egocentric/objective distinction, as applied to temporal representation, also provided the opportunity to introduce a metaphysical debate that will feature very prominently indeed in this book: namely, the debate over whether time passes. According to the A-theorist, time does pass as a matter of objective fact, and this is reflected in our used of tensed expressions—those, that is, that intimate position as past, present, or future. Apparently non-tensed expressions are parasitic on tensed ones, and this reflects the supervenience of facts about earlier/later relations on facts about past, present, and future. According to the B-theorist, in contrast, time does not flow, and we can coherently represent time in terms that do *not* intimate position as past, present, or future. The A-series is mind-dependent, *not* in the sense that only mental events instantiate A-series position, but in the sense that the past/present/future distinction is confined to representation. The metaphysical issue is thus bound up with the nature of temporal representation: what metaphysical view we take of time will have consequences for our view of how our temporal representations map on to reality. Equally, views on temporal representation may constrain our view of the coherence of either side of the A-theory/B-theory debate. In particular, if we think that our understanding of B-series expressions like 'earlier than' are parasitic on tensed, A-series expressions like 'present', then we will regard the B-theory, which offers to give tenseless B-series truth-conditions of A-series statements, as either incoherent or circular. Given the force of that objection to the B-theory, B-theorists cannot simply help themselves to the assumption that a tenseless language is available in which the truth-conditions for tensed beliefs and judgements can be articulated. What needs to be shown is that there is a role for such a language. Relatedly, the B-theorist also needs to show that we have available to us a purely B-theoretic representation of time. One case that illustrates the need for such a way of thinking of time is the subject of the next chapter: memory.

PART II

MEMORY AND PERCEPTION

4

Retracing the Past: Memory and Passage

> But Time, to make me grieve,
> Part steals, lets part abide;
> And shakes this fragile frame at eve
> With throbbings of noon-tide.
>
> Thomas Hardy, 'I look into
> my glass'

4.1 PRINCIPLES OF EPISODIC MEMORY

I am trying to find a book I have mislaid. It is not in its usual place on the bookshelf. It is not on the bedside table, or on my desk. Frustrated, I set about making a pot of tea. Suddenly, an image comes to mind: I picture myself handing the book to a friend, recommending it as a good read. 'No hurry about giving it back', I hear myself saying in my mind's ear, 'I shan't need it for some time'.

I remember, then, lending the book to a friend. But I do not merely remember *that* I lent it to them. I remember the experience of doing so, what the book looked like, the expression on my friend's face, the sound of my foolish utterance. What we have here is an example of what psychologists since Tulving have called *episodic* memory: remembering an event 'from the inside', recalling the experience itself (see e.g. Tulving 1983). It is often distinguished from remembering that, or *semantic* memory: for example, remembering that Schliemann's discovery of what he took to be the mask of Agamemnon occurred in 1876. What is required for an episodic memory of an event, but not for a semantic memory concerning it, is that the event in question was experienced by the rememberer. Of course, the semantic memory will have arisen from some relevant experience, but it does not need to have been a perceptual experience of the event itself, as opposed to hearing or reading a report of it.

Does the episodic/semantic distinction capture an objective difference between kinds of memory, or is it merely conventional? We will consider reasons for thinking it conventional later on, but first I want to try to capture the notion of episodic memory more precisely, paying particular attention to its epistemological features.

Of these two kinds of memory (allowing the distinction for a moment), it is episodic memory that seems to link us most intimately with the past. We can, in fact, characterize the difference between episodic and semantic memory as that between knowledge by acquaintance and knowledge by description, where the object of knowledge is the past. It is tempting, indeed, to describe episodic memory as a *re-experiencing* of the past. That description should not be taken too literally, but it is a feature of episodic memory that it is not just a representation of a past event; it is also a representation of the *experience* of that event. So here is my proposed definition:

An episodic memory of an event (i) arises from an original experience of that event and (ii) includes that experience (or a representation of that experience) as part of its content.

In what follows, I am not going to pay much attention to the second component of that definition, concentrating rather on the part of the content of the memory that concerns the event itself.

The question I want to ask is this: Does the link between episodic memory and the past tell us anything about the nature of time? I think it does.

The argument starts with an epistemological principle concerning episodic memory. It is this: a truly episodic memory cannot acquire a greater degree of closeness to the truth than the original experience. The memory is accurate only to the extent that the original experience was accurate. Call this principle the *Accuracy Constraint*. (Just to clarify: I am treating experience and memory here as items with propositional content, and so as vehicles for truth and falsity, not simply as a collection of sensations.) Now we might think that it would be quite easy to come up with counterexamples to this principle. Suppose, walking in the park one day, you are an unseen witness to what is evidently a very emotionally charged meeting between two people. You recognize these two people as friends of yours: a married couple. Seeing evidence of intense affection between them, but also what seems to be anxiety, you do not announce yourself, but walk quietly away. Later that day you learn, quite fortuitously, that the woman you thought you had seen in the park is in fact in some remote location, visiting an aunt. In an instant you recognize the significance of what you saw: not a meeting between husband and wife, but . . . *an affair*! You now see the scene with different eyes (or different mind's eyes). You see the woman, but not as the woman you know. Perhaps you now see her as younger, taller, with darker hair. Is this not still an episodic memory? The experience is re-presenting itself, but it is now a reinterpreted experience. The memory, it seems, has become closer to the truth than the original experience.

This is an entirely coherent story, but it does not, I think, impugn the Accuracy Constraint. The additional information turns the memory into a rather different kind of memory, no longer a purely episodic memory of the original experience. You might now report the memory like this: 'I recall seeing two people, and thinking at the time that they were a married couple of my acquaintance. I now know that one of them was not the person I took her to be.' The genuinely episodic element of the memory is not changed by the later information, only its significance is.

Is there any limit to the extent to which your original interpretation of what you saw could be revised? What if you subsequently discover that you misidentified the man, too, so that your assumption of an affair was quite unwarranted? Or perhaps the two people in question were rehearsing a play, or were not people at all, but incredibly sophisticated robots, or holograms. . . . If almost everything about the original experience is subject to possible revision, then is the core component of the episodic memory—the one that is preserved in the memory—genuinely propositional at all, or simply a collection of qualia? Doubtless we can imagine a process like this in which the content of the original experience that is preserved in later judgements is gradually pared down. But it does not follow that episodic memories are not propositional. An experience does not have to be veridical in order to be preserved in episodic memory. A memory of an experience that turned out to be completely illusory is no less a genuine memory of that experience. The point, then, is that although our interpretation of an experience might force us to abandon our original experiential belief, whatever is *preserved* in episodic memory of an experience cannot be closer to the truth than the original experience, because any beliefs you subsequently come to have are, by definition, not part of the original experience. And if they are not part of the original experience, then they are not part of the episodic memory, but rather of a larger set of mental states that includes that memory.

Another kind of apparent counterexample is constituted by cases where recalling an experience, even in the absence of subsequent information, reveals information that was not available to the subject at the time of the original experience. In response, we can certainly concede that the process of recall may lead to beliefs that do not correspond to any conscious belief held at the time of the experience. But the content of those later beliefs must have been encoded in some form in the original experience. What this suggests is that we should widen the concept of the 'original experience' to include sub-doxastic states.

As my remarks on the meeting in the park case will have suggested, what I am proposing is that the Accuracy Constraint is less a substantive fact about episodic memory than something that simply follows from our definition of it. The reason why episodic memories cannot be more veridical than the original experience, I suggest, is that the content of the episodic component of a memory is determined by the memory's link to the original experience, and in addition that the *well-groundedness* of the memory derives from that of the original experience. Call this principle the *Inherited Ground and Content Principle*. The 'content' part of the principle asserts that the representational content, and hence closeness to the truth, of that component is what it is by virtue of the content of the experience, and contains nothing that was not part of the content of the experience. (Of course, given that memories fade, the representational content of the memory may be somewhat less than the content of the experience. But it cannot be more.) So the Inherited Ground and Content Principle entails the Accuracy Constraint.

Something quite close to the Inherited Ground and Content Principle appears in John Campbell's (1994) discussion of memory. As he puts it, memory, like testimony, is not an independent channel to the truth. Memory could not be the only source of knowledge: it depends on there being another source, namely perception. Campbell distinguishes between 'ground-floor' and 'reflective' conditions on knowledge. Ground-floor constraints include appropriate causal connectedness with the state of affairs that is the object of knowledge, whereas reflective conditions would include subjects' own views on the process by which they formed their beliefs. Campbell's principle governing episodic memory is this: a memory cannot count as knowledge unless the original experience met the ground-floor conditions (though it may not have met the reflective conditions). This he calls the 'stepwise' conception of memory (Campbell 1994: §7.3).

Once the Accuracy Constraint is made immune from counterexample in the way I have suggested, we can continue to employ the episodic/semantic distinction in the face of warnings from psychologists that the distinction is largely conventional. Here, for instance, is the psychologist Alan Baddeley's verdict:

An example of semantic memory might be knowing the chemical symbol for salt, while episodic memory would be exemplified by remembering a personally experienced event, such as meeting a retired sea captain while on holiday. There clearly *are* differences between these two situations, but it is questionable whether a distinction based on anything as subjective and phenomenological as personal reference is either viable or appropriate. Since all memory is surely based ultimately on personal experience, it is hard to see what is gained by assuming different memory stores depending on whether the personal reference is or is not recalled.

An alternative way of conceptualizing the difference between remembering personal incidents and recalling information is in terms of the degree of abstraction involved. . . . long term memory has a strong abstractive component; we tend to minimize memory load by stripping away inessential details and encoding new material in terms of existing schemata, keeping only enough to allow us to reconstruct the event if recall is required. . . . It is . . . not only the case that semantic memory is built from personal experience by a process of abstraction; it is also the case that what appears to be a direct record of personal experience is itself a reconstruction based on an abstraction. . . . Hence, although Tulving's distinction between episodic and semantic memory may provide a useful reminder of the range of semantic memory, it remains very doubtful whether this reflects a clear dichotomy between separate storage systems, as Tulving suggests. (Baddeley 1976: 317–18)

What these remarks show is that, for psychologists, the issue is about encoding mechanisms. But philosophers, who tend to operate at higher levels of abstraction, can remain agnostic about this. What I have suggested is that, in so far as we are concerned with content and causal history, the distinction remains entirely viable and appropriate. The episodic element of a memory is simply defined as that to which the Accuracy Constraint and the Inherited Ground and Content Principle apply. These principles are admittedly somewhat more demanding than Campbell's stepwise conception of memory, but then, as he points out, the stepwise conception applies equally to many cases of semantic memory.

Once the Inherited Ground and Content Principle is in place, a further thesis emerges, one about truth-makers: whatever in reality made true, or veridical, the original experience is necessarily connected to whatever makes true, or veridical, the later memory. Call this the *Truth-Maker Connection*. What is this necessary connection? The simplest answer we can give to this is that experience and memory have exactly the same truth-maker. The picture is unlikely to be quite as straightforward as this, however, as the memory, unlike the original experience, represents the event as *past*. This is not adding to the content of the memory in such a way as to come into conflict with the Accuracy Constraint, for if the purpose of memory is to preserve information about the past, then it can do so only by truly reflecting the fact that time has passed since the event took place. Passage from 'this is *now* happening' to 'this happened' is therefore required if the memory is to be as close to the truth as the original experience. But the experience and the memory will still have something in common in terms of content. What I propose, then, is that the truth-maker of the experience is a component of the truth-maker of the memory. This then guarantees that the memory can be true only if the experience was true.

So we now have a number of connected principles characteristic of episodic memory:

The Accuracy Constraint: The memory is no more veridical than the original experience.

The Inherited Ground and Content Principle: The content and well-groundedness of the memory are determined by those of the original experience.

The Truth-Maker Connection: The truth-maker of the memory is necessarily connected to that of the original experience.

The second of these principles, I have suggested, implies the other two. I am going to argue that these principles, and especially the second two, motivate a particular conception of time introduced in the previous chapter: namely, the B-theory. Let us begin by taking a closer look at the Truth-Maker Connection, and inquire into the nature of the relevant truth-makers.

4.2 TWO MODELS OF MEMORY, AND A PUZZLE CONCERNING TRANS-TEMPORAL ENTAILMENT

Consider the event I recalled earlier of my lending a book to a friend. Is that event receding into the past or not? To recall our discussion in the last chapter, we can think of events in time as constituting the A-series, that is, as being more or less past, present, or more or less future—positions which, of course, would always be changing. Or we can think of events as constituting only the B-series, that is, as standing in the unchanging relations of temporal precedence or simultaneity to each other. Now, if there really is an A-series, if time really is passing, then B-series relations are simply a product of A-series position. This is the A-theory. If, for instance, x is past and y is future, then x is earlier than y. To hold that there is an A-series is not to assert the unreality of the B-series. But there is no plausible way to invert this and represent the A-series as supervening on the B-series in such a simple way. Passage does not simply emerge from order. Those who hold that the B-series does *not* supervene on the A-series typically hold that the A-series is unreal, that time does not pass: there is only temporal order. This is the B-theory. It would, perhaps, be possible to hold that A-series passage and B-series order were completely independent of each other, but this would be a deeply unattractive position.

So those who, for whatever reason, hold that events form only a B-series, have something to explain: when in ordinary speech we appear to allude to the A-series, as when we say 'I'm flipping the switch *now*', or 'My aunt arrived *yesterday*', or 'The lunar eclipse will occur *tonight*', do we speak truly? It is here that the metaphysician of time has to engage with the semantics of tensed discourse. Again, as we saw in the last chapter, on one account (there are others), if I say 'I'm flipping the switch *now*' at time t, where t is a position in the B-series, then what I say is true if and only if it is a fact that I flip the switch at t. This semantics, and the associated assertion that there is in reality no A-series, is the *date B-theoretic semantics*.

Contrasted with this is what we might think of as the natural semantics for tensed statements: 'I'm flipping the switch *now*' is true if and only if I am flipping the switch now, in the objective present. This semantics, and the associated assertion that A-series terms reflect corresponding divisions in reality, is the *simple A-theoretic semantics* (we shall consider the date A-theoretic semantics shortly). A-theorists do not always want to be associated with the idea of events literally receding in time, as if pastness is something an event can acquire more of. Some A-theorists, indeed, are *presentists*: they hold that only what is present is real (see Bourne 2006 for a detailed articulation and defence of this view). Outside the present, there are for them no events to *be* past. Believing in the A-series is for them a matter of 'taking tense seriously', to use a phrase of Ned Markosian's, that is, taking the tensed structure of language to be primitive and irreducible (Markosian 2002). And what that requires is a semantics of tensed discourse and belief that reflects that irreducibility. In what follows, I shall be concerned precisely with such a semantics, rather than a literal picture of events as receding into the past. The argument will be neutral, however, on the issue between presentist versus non-presentist interpretations of the A-theory.

The crucial difference between the A-theory and the B-theory is their view of the kinds of thing that make temporal statements and beliefs true: namely, the *facts* of the matter. Here I am taking 'facts' to be constituents of reality. For the B-theorists, facts do not change. For the A-theorist, they do. The facts which are constitutive of the present—for instance, that you are now reading this book, that the television is on as you are doing so, that Mars is such-and-such a distance from the Earth, etc.—are in due course replaced by other facts—for instance, that you are no longer reading this book but giving your full attention to the television.

Corresponding to these two views of reality are two models of memory. For the A-theorist, our changing beliefs about a given event track the changing facts of reality (the A-model). Your belief that a certain event—for example, the storm now raging outside—will in due course be replaced by the belief that it is over. On this model of memory, the mind tracks *changing* states of affairs, viewed from the *same* perspective (the present). What we have, in effect, are two parallel processes: the external changing facts and the internal changing beliefs, which run in pre-established harmony (that is, evolutionarily established harmony) (see Figure 4.1).

Fig. 4.1: The A-model of memory. The dotted lines here represent some natural process by which some item is replaced by another. The continuous line represents the aboutness relation between a mental state and its object.

Fig. 4.2: The B-model of memory.

For the B-theorist, in contrast, our changing beliefs about an event do not track *changing* facts, but rather the same fact from different perspectives (the B-model) (see Figure 4.2). The contrast between the A- and the B-model of memory can be illustrated by means of a spatial analogy. Imagine a perceiver walking around a fixed object, but keeping it in view all the time: the same object is viewed from different perspectives. This would be the spatial analogue of the B-model. Contrast this with a perceiver who does not move, but is presented with a series of different objects in different orientations: here, different objects are viewed from the same perspective. This would be the spatial analogue of the A-model.

The B-model, I am going to suggest, turns out to have the advantage when we consider again the principles governing episodic memory.

As I handed the book to my friend, I had an experience of seeing the book first in my hand and then in hers. I heard myself say 'No hurry about giving it back', and saw her smiling in response. I formed various present-tensed beliefs, among them 'I am lending the book to her'. Later, recalling the experience, I think 'I *lent* the book to her'. The later belief can only be true if the earlier one was. That is what the Accuracy Constraint requires.

What we have here is an example of what is called 'truth-value links': a necessary connection between the truth-values of judgements made at different times. A somewhat more transparent description of the phenomenon would be *trans-temporal entailment*.

How does this fit with the two accounts of the semantics of tensed belief, the A-theory and the B-theory? On the B-theory, what made my original perceptual belief at *t*, 'I am lending the book to her', true, is the fact that the lending takes place at *t*. This same fact is part of the truth-maker for my later memory belief at *t**, 'I lent the book to her'. The other part of that truth-maker is the fact that *t* is earlier than *t**. The truth of the memory therefore entails the truth of the perceptual belief.

The Accuracy Constraint is thus satisfied: the closeness to truth of the memory belief depends on the closeness to the truth of the perceptual belief: if the earlier belief had been false, the later belief could not have been true.

What of the A-theory? According to this account, the truth-maker of the original belief is that I am *now* lending the book to my friend. By the time I recall the incident, however, this fact has gone out of existence, to be replaced by the fact that I *lent* the book to my friend. The past fact, the one that made true my original belief, is no longer available to make any present belief true. It cannot even be a component of the truth-maker of the memory. My present memory belief is made true by a *present* fact, that I lent the book. Original belief and later memory are therefore, on this account, made true by *different* facts.

Does this matter? It does if this means that we no longer have an explanation for trans-temporal entailment. If the original belief and later memory have different (indeed, incompatible) truth-makers, why should the truth of the memory depend in any way upon the truth of the original belief? The A-theorist may reply that, although the truth-makers are quite different, the obtaining of the later truth-maker, that *I lent the book*, nevertheless necessitates the past obtaining of the earlier truth-maker, that *I am lending the book*. This connection would then ground the relation of trans-temporal entailment between the beliefs: the truth of the later belief entails the truth of the earlier belief, and vice versa. Unfortunately, the A-theorist cannot express the required connection between the truth-makers. One of the consequences of 'taking tense seriously' is that every attempt to express the relationship between facts existing at different times ends up expressing a relation between present facts. Consider this report:

If my present belief 'It *was* the case that *p*' *is* true, then my past belief 'It is *now* the case that *p*' *was* true.

This certainly looks like an expression of the relationship of trans-temporal entailment: a truth at a later time entailing a truth at an earlier time. But this whole sentence is true *now*, and the truth-maker of both antecedent and consequent are present facts. As long as we think that truths *about* time can be stated only from the perspective of a position *in* time (namely, the present), then we will never be able truly to express the relationship between a fact that obtains at one time and a fact that obtains at another time.

Suppose, however, that the A-theorist responds by pointing out that the truth-makers of token beliefs change over time, so that what made true my original experiential belief 'I am *now* lending the book' at the time I had it is not what makes that same belief true at a later time. The A-theorist can appeal to what in the previous chapter was called the *date A-theoretic semantics*, according to which the content of a token belief changes with the context of assessment:

A token belief, 'I am lending the book', tokened three days ago, is (now) true if and only if I lent the book three days ago.

Now this truth-maker will *also* serve as the truth-maker for my current memory belief, 'I lent the book', so now truth-maker for experience and memory is one and the same, and the trans-temporal entailment relation is secured. We no longer have to connect the truth-maker that the original experiential belief had at the time it was tokened with the truth-maker that my current memory belief has.

This would indeed solve the problem if it were just about trans-temporal entailment. But there is another dimension, and that is the causal story behind my acquisition of the memory. Consider the Inherited Ground and Content Principle: the content and well-groundedness of the memory are determined by those of the original experience. The process by which groundedness is inherited is surely causal: the grounds of the original experience are part of the causal history of the later memory. Moreover, according to the Causal Truth-Maker Principle introduced and defended in Chapter 2, the truth-maker of the memory must itself appear in that causal history. Putting the CTMP together with the conception of episodic memory articulated in this chapter, we have the following account:

(a) If the original experience is sufficiently well grounded to count as knowledge, then its truth-maker must appear in the causal history of that experience.

(b) The later memory inherits the well-groundedness and content of the original experience by virtue of the fact that the truth-maker of the original experience also features in the causal history of the memory.

(c) The truth-maker of experience and memory must necessarily be connected in such a way as to explain the trans-temporal entailment relations between them.

Can the A-model satisfy all these requirements? We have just seen that it can satisfy (c) by adopting a suitable semantics for tensed judgements: both original experience and memory currently have the same truth-maker. However, this manœuvre prevents the account from satisfying (a) and (b). If the original experience always has a *present* truth-maker, however far that experience has receded into the past, how can that truth-maker feature in the experience's causal history?

The argument against the A-model (and therefore against the A-theory) may be summed up as follows:

(1) Episodic memory requires a connection between the truth-makers of memory and the experience on which the memory is based, truth-makers that feature in the causal history of those states.

(2) That connection cannot be articulated if temporal truths are all present truths.
    Therefore:

(3) Only the B-theory can provide a satisfactory account of episodic memory.

4.3 A PARALLEL PROBLEM: ANTI-REALISM ABOUT THE PAST

Before going on to defend this argument against objections, I want to consider a precisely parallel difficulty for a certain view of the past.

The existence of truth-value links between statements made at different times (what I am calling 'trans-temporal entailment') and their significance is considered in Michael Dummett's well-known paper 'The Reality of the Past' (1969). The central question of that paper is whether these links pose a problem for *anti-realism* about the past, the view that past-tense statements are made true by evidence which justifies the statement. The difficulty, as he acknowledges, is that judgements made in the past about what obtained then and corresponding judgements made in the present about the past will on the anti-realist view have very different truth-conditions. Suppose at one time I say, truly, 'There are fairies in my garden'. But by the time I come to say, at a later date, 'There were fairies in my garden', all trace of their existence has been completely obliterated. The later statement will, on the anti-realist view, be either false or lack truth-value. The truth-value link from the earlier statement to the later one has thus been broken. We might also imagine the following, perhaps rather less usual case: I utter 'There are fairies in my garden' when there is no evidence warranting the assertion at all. On the anti-realist view, the utterance lacks a truth-value. But now suppose that God (for example) creates evidence warranting the later assertion, 'There *were* fairies in my garden'. This later statement is true, but there is, bizarrely, no truth-value link with the earlier statement.

Dummett's reply on behalf of the anti-realist, is, in effect, that the assertion of a truth-value link will itself be an assertion that takes place in time, and so be evaluable in the context of that time. If I now have evidence that justifies me in asserting 'There *were* fairies then', then I also have evidence which justifies me in asserting 'If I said then "There are fairies", I would have spoken truly'. To demand, as the realist does, that we ignore our temporal perspective in discerning the truth-value link is to demand the impossible (Dummett 1969: 368–70).

Campbell, however, argues that this suggestion does not allow us to preserve the stepwise character of memory, such that, as he puts it, 'The groundedness of the memory judgement depends on the groundedness of the original judgement' (Campbell 1994: 233). Part of the connection here, of course, is the truth-value link, and again we get an apparent conflict with the anti-realist view of the past. But here the anti-realist's explanation of the truth-value link is inadequate. My current memory judgement is true by virtue of current or future evidence in justification of it. But it is not *this* evidence to which the original judgement was sensitive. The value of bringing memory into the discussion is this: even if anti-realism is consistent with truth-value links (trans-temporal entailment), its account of these links is not consistent with the idea that in memory we are tracking the same state of affairs as occasioned the original perceptual judgement.

What the argument of §4.2 does is to employ essentially this move (somewhat more formally expressed) against the A-theory. The A-theorist, as characterized here, need not be an anti-realist about the past in Dummett's sense. The passage of time may take things and events out of existence, but it does not follow that statements about them are true only in virtue of evidence we are capable of finding. Nevertheless, for the A-theorist, as for the anti-realist, the facts of the matter are constantly changing (though possibly the anti-realist would object to this way of putting it). The facts to which the original perceptual experience was sensitive are not the facts against which the truth of the memory is to be assessed. More seriously, the A-theorist's ontology prevents him from articulating just what the connection is between those facts.

Here is Dummett on the crucial difference between the realist and the anti-realist:

What the realist would like to do is to stand in thought outside the whole temporal process and describe the world from a point which has no temporal position at all, but surveys all temporal positions in a single glance. . . . The anti-realist takes more seriously the fact that we are immersed in time: being so immersed, we cannot frame any description of the world as it would appear to one who was not in time, but we can only describe it as it is, i.e. as it is now. (Dummett 1969: 369)

That also captures the difference between two brands of realism: the A-theory (immersed in time) and the B-theory (standing outside it). And standing in thought outside time is precisely what we would have to do in order to articulate the connection between a fact existing in the past and giving rise to the original experience and a fact existing in the present and making true my current memory.

4.4 SOME OBJECTIONS

I will conclude this chapter by considering three objections to the argument against the A-model.

First, it might be argued that in fact the A-model and the B-model are in exactly the same position vis-à-vis the explanation of trans-temporal entailment. For both models can allow that the truth-makers for the two beliefs have a component in common. On the B-model, the common component is the fact that I lend the book at t. But this is not sufficient for the truth of the later memory belief: an extra component is required, as the following schema shows.

The belief 'I am lending the book', had at t, is made true by the fact that:
(i) I lend the book at t.

The belief, 'I lent the book', had at t^*, is made true by the facts that:
(i) I lend the book at t; (ii) t is earlier than t^*.

Fact (ii), that t is earlier than t^*, is a component of the truth-maker of the memory, but not of the truth-maker of the original experience.

Similarly, on the A-model, there is (so the suggestion goes) a common component to the truth-makers of original experience and memory, namely an *event*, that of my lending the book. But this is sufficient for the truth of neither belief: what more is required is a fact about the temporal location of this event. Let us say that we go for a simple A-theoretic semantics:

The belief 'I am lending the book' is made true by:
(i) the event of my lending the book; (ii) that event's occurring *now*.

The belief, 'I lent the book' is made true by:
(i) the event of my lending the book; (ii) that event's occurring *in the past*.

(Of course, this analysis would not be available to the presentist, for whom the event in question has ceased to be part of any reality that could make the memory belief true.)

Is this an effective response? I suggest not, for there is this crucial difference between the A-model and the B-model: on the B-model, the *complete* truth-maker for the original experiential belief is a part of the complete truth-maker for the memory belief. So if the memory belief is true, the original experiential belief is true. On the A-model, the existence of the complete truth-maker for the original experience is actually incompatible with the existence of the complete truth-maker for the memory. So the A-model's representation of the truth-makers, given the simple semantics, cannot explain the trans-temporal entailment relations between original experience and later memory.

What, then, if we move to a more complex A-theoretic semantics, according to which the truth-conditions for a belief vary over time? Then we have the following:

The belief 'I am lending the book', had at some past time t, is made true by:
(i) the event of my lending the book; (ii) that event's occurring at t.

The belief, 'I lent the book', had now, is made true by:
(i) the event of my lending the book; (ii) that event's occurring earlier than now.

The truth-makers of the experiential belief guarantee the truth of the memory belief, since if e occurred at some past time, it also occurred earlier than now. But although trans-temporal entailment is accounted for, this falls foul of the Inherited Ground and Content Principle, which (on a causal account of inheritance) requires the truth-maker for the experiential belief to be part of the causal history of the memory belief.

The second objection tackles the causal problem: how can experience and memory share a causal history? Instead of regarding causes as facts, we could think of them as events. Now, unless we are presentists, events do not cease to be real once they become past. So causal relations can hold between non-contemporaneous events. But since, according to the A-theorist, time flows, so that one set of facts is replaced by a different set of facts, those events and causal relations are constituents of different facts over time. This seems to be a perfectly coherent story to tell about causation. Unfortunately, it does not fit with the Causal Truth-Maker Principle, for then the truth-makers of experience and memory (which involve more than just the event in question) cannot be their causes.

The upshot of our consideration of these two objections is that, although the A-theorist can tell a coherent story about truth-makers (and trans-temporal entailment) *and* a coherent story about causation, the two stories do not combine in the way that they need to if we are to understand the epistemology of memory. Putting the phenomenon of trans-temporal entailment in the context of memory thus adds an extra dimension: a causal dimension. Experience and memory are well grounded by virtue of their causal history, which should include their truth-makers if they are to warrant the title of knowledge. Even if we are prepared to accept that the original experience is now made true by a present fact, it makes no sense to suppose that it was *caused* by that present fact.

Thirdly, and finally, we should consider an objection to our original account of episodic memory as characterized by the three principles (and to Campbell's stepwise conception of memory). The suggestion that our access to the past through memory is indirect, so the objection goes, is unwarranted, and is just as problematic as the corresponding theory of perception. Let us develop this in some detail.

The philosophy of perception in the last hundred years has been dominated by the debate between two theories (or variants of these theories): the *direct*, or 'direct realist', theory and the *indirect*, or 'representative realist', theory. According to the direct theory, the objects of our perceptions of external things are the things themselves. We just see the tree, hear the river, touch the flower. According to the indirect theory, in contrast, the immediate objects of perception are items that are wholly mind-dependent, items that are sometimes described as 'sense data' (Broad 1923; see Robinson 1994 for a more recent defence). These sense-data represent the external objects, which we therefore can be said to perceive mediately. For both direct and indirect theories, the ultimate objects of perception exist quite independently of our perception of them (which is why they are sometimes referred to as 'realist' theories). Both, therefore, reject the idealist or phenomenalist contention that external objects (or statements concerning them) are somehow constructions out of (made true in virtue of) our mental states. The difference between direct and indirect theories lies in whether the *immediate* objects of perception are mind-dependent or not. Part of the motivation for the indirect theory is provided by reflection on cases of hallucination: if, when I am hallucinating, there is *something* that I see, the object of my perception can only be the sense-datum: there is nothing external to me that constitutes such an object. Direct theorists would reject the crucial assumption: why should there be an object at all, rather than merely a (mistaken) way of representing the world? For the direct theorist, we can talk about how the world appears, without reifying appearances.

Since episodic memory is in some ways analogous to perception, it is natural to ask: Is there a corresponding debate about episodic memory? Or perhaps we should ask, is there a correspondingly important debate about memory? For it may be that, although we can construct counterparts to the two theories, one of them may appear platitudinous, or otherwise undeniable, and the other absurd.

The indirect theory of memory, then, asserts that remembering involves entertaining or contemplating a memory image. The immediate object of thought is this image, representing some past state of affairs which thus constitutes the mediate object of thought. We recall some past episode only by virtue of recalling something else: the memory image. (Compare: we perceive the tree only by virtue of perceiving the corresponding sense-datum.) The direct theory of memory would simply deny this: whether or not we can talk of memory images, we do not recall a past event by recalling something else (the image): we simply recall the past event. Remembering is simply direct access to the past, a view we find in Bergson 1912; Laird 1920; and Russell 1912 (though contrast his 1921: ch. 9).

If the direct theory of memory is correct, then we should, after all, regard memory as an independent channel to the truth: it does not inherit its well-groundedness from the well-groundedness of an original experience. But if this is so, then the whole argument of §4.2 is undermined.

Or is it? Memory, of course, is not perception, even if there are parallels. Even the direct theorist would not want to say (would they?)

that in memory we are directly experiencing the past in a quasi-perceptual manner. So what exactly does 'direct access' mean in this context?

One possibility is direct *causal* access to the past: that is, that there is no causal intermediary between the memory and the past event it represents. But this raises two problems. The first is that it seems to involve unmediated action at a distance. The second, a point made by Campbell, is that we then lose what would otherwise be a natural explanation of why our episodic memories are only of events that took place when we were physically present. If an episodic memory is causally linked to a prior experience, and gains its content from that experience, then there is no mystery here. We can only have experiences of things when we are appropriately physically located. But once the tie between memory and original experience is cut, this natural explanation is no longer available to us. To put it another way: we would ordinarily take evidence that we were not in a certain place at a certain time to undermine our claim to have an episodic memory of what happened there and then. But on this construal of the direct theory, such evidence should have no bearing on the matter (Campbell 1994: 237–8).

To reinforce the point: the spatially perspectival aspects of an episodic memory are naturally explained by our spatial orientation at the time of the original experience. What less would explain them?

Another possible interpretation of 'direct access' is direct *epistemological* access: knowledge of the past is not inferred from some memory image, but rather the memory provides me with knowledge of the past directly (a view defended by Don Locke (1971)). But this is entirely consistent with the epistemological principles we introduced in the previous section: the memory simply inherits the well-groundedness of the original experience; it does not provide *different* grounds for belief. We might also note in passing that if this is what the direct theory of memory amounts to, then it is not the precise analogue of the direct theory of perception. One could be an indirect theorist of perception to the extent of believing in the existence of sense-data as the immediate objects of perception, whilst still holding that they provided non-inferential knowledge of the present.

Finally, access to the past could be direct in the sense of not involving memory imagery. But if this is an issue about encoding mechanisms, then it seems to be more of a psychological than a philosophical issue, and again it implies no conflict with the principles of episodic memory articulated in § 4.1.

In this chapter I have suggested that the B-theory sits much more comfortably with our intuitive view of the epistemology of memory than does the A-theory. But there is another aspect of our experience that inclines us to favour the A-theory: namely, perception. And the support that perceptual experience gives to the A-theory seems to be very much more direct than the rather complex considerations adduced in this chapter. To perception, then, we now turn.

5

Projecting the Present: The Shock of the Now

If a man stands before a mirror and sees in it his reflection, what he sees is not a true reproduction of himself but a picture of himself when he was a younger man.

Flann O'Brien, *The Third Policeman*

5.1 THE ARGUMENT FROM EXPERIENCE

Why does the A-theory strike us as the natural view of time? What is it, pre-theoretically, that inclines us to suppose that there is something special, and not merely perspectival, about the present? Why do we say, and think, that time passes? Surely the answer lies in our ordinary perceptual experience. We describe things as happening *now* when we perceive them happening. We do not (ordinarily) talk of perceiving the future or the past. There are, in other words, temporal limits to our perception, and as these limits are not chosen by us, we naturally suppose that it is some feature of the world that narrows our attention to this moment of time we call the present. And the simplest way of articulating this is to say that it is because events *are* present that we see them as such—indeed, that we see them at all. As for the passage of time, we are not only aware of this when we reflect on our memories of what has happened. We just *see* time passing in front us, in the movement of a second hand around a clock, or the falling of sand through an hourglass, or indeed any motion or change at all. This is a feature made much of by A-theorists. Consider the following remarks:

There is not only a diversity in the contents of our acts of awareness from time to time but also these acts of awareness themselves appear to undergo becoming.

If these appearances are to be saved, it must be granted that mental events are intrinsically past, present and future and change in respect of these A-determinations. (Gale 1968: 233)

Let me begin this inquiry with the simple but fundamental fact that the flow of time, or passage, as it is known, is given in experience, that it is as indubitable an aspect of our perception of the world as the sights and sounds that come in upon us, even though it is not the peculiar property of a special sense. (Schuster 1986: 695)

There is hardly any experience that seems more persistently, or immediately given to us than the relentless flow of time. (Schlesinger 1991: 427)

If time did not pass in reality, what would explain our perception of this constant succession of events? For many, if not all, A-theorists, the phenomenological data of temporal experience are entirely explicable in terms of the A-theory, but either inexplicable in terms of, or actually incompatible with, the B-theory. Let us call this 'the argument from experience'.

Calling this '*the* argument' is perhaps misleading. What we have is really a series of arguments, each starting from a different phenomenological datum. Here is a list of the relevant data:

- We perceive events as present, i.e. as happening *now*.
- We perceive only the present.
- Our experience is temporally limited.
- In normal perceptual conditions, similarly spatially located observers agree, by and large, on what is happening now.
- Our experience is constantly changing.
- We perceive change and motion.

Clearly, some of these are more purely phenomenological than others, and some imply more about the world than others. The second, in particular, has metaphysical overtones that threaten to trivialize the argument from experience. In what follows, the aim is to show, for each of the data above, that it is entirely compatible with, indeed explicable in terms of, the B-theory. In the last section of the chapter, I consider evidence to suggest that our perception of what we ordinarily take to be the present is far from passive, but involves an active projection of states or events on to the world.

5.2 THE EXPERIENCED AND SPECIOUS PRESENTS

Take the first datum, 'We perceive events as present.' This seems undeniable, but it is not clear that we are being presented with a significant fact about experience that requires explanation. What is 'perceiving something as present' being contrasted with? Perhaps perceiving something as past. But it is not clear that we can make sense of this. Of course, we might hear some distant event (a thunderclap, for instance), and, realizing that sound only travels at finite speed, believe that the event itself happened a while ago (especially if we saw the associated lighting flash a few seconds ago), but this is not to hear it *as* past. An event we believe to be past does not have a special sound or visual appearance (Mellor 1998: 16).

Russell once suggested (1921) that our memories, and more particularly our episodic memories, are accompanied by a characteristic feeling of pastness. But this is clearly not a case of perceiving something as past, since it is not a case of perceiving at all, unless one thinks that we perceive our memory images, but then the *image* will be taken to be present. Such a feeling of pastness seems to be nothing more than a belief that the event in question is past.

It is not clear, then, that there is an interesting difference between perceiving something 'as present' and simply perceiving it. In which case 'We perceive events as present' tells us nothing more than that we perceive events. So the first datum has not identified any particular aspect of perception that requires explanation. (It may be true that the mere fact that we perceive things ultimately requires appeal to the A-theory, but it is not at all obvious how an argument for this would proceed, if not by identifying a particular aspect of experience.)

It is, to be sure, an interesting feature of perception that perceptions are accompanied by present-tense beliefs, that for instance seeing the thunderstorm is accompanied by the belief that the thunderstorm is going on now. This is a logically contingent (though biologically necessary) feature of experience, and perhaps represents the truth behind the remark that we experience things as present. But this fact seems entirely neutral as far the A-theory/B-theory debate is concerned. If it requires explanation, the explanation is more likely to be a causal one—that forming the appropriate beliefs in response to perception is very obviously a trait favoured by natural selection—than one that involves essential appeal to the metaphysics of time.

Now consider the second datum, 'We perceive only the present.' This certainly does not seem to be theoretically neutral. Indeed, if we take it to imply a non-perspectival division of reality into present and not-present, then it is clearly not compatible with the B-theory. But, so interpreted, it is false. Information from external objects, in the form of light and sound, takes time to reach us. Once it has arrived, the information has to be processed by our sensory systems, and this takes time. The upshot is that by the time we have registered an event or state of affairs, it will have receded into the past. So, at best, we perceive only the very recent past. (And, in the case of very distant objects, such as stars, it is the distant past that we perceive.)

There is a further discrepancy between the experienced present (i.e. what we experience as present) and any supposedly objective present. As Augustine clearly saw, any objectively present moment would have to be durationless. Here, in a nutshell, is the argument. Suppose the present to last for a non-zero interval. It would then have to be divisible into earlier and later parts. But if it is so divisible, then its parts cannot all be present. If some earlier part is present, then some later part is future. Or, if some later part is present, then some earlier part is past. Therefore, the present must not be divisible in this way. Therefore, it must be durationless (*Confessions*, Book XI, §15; Pine-Coffin 1961: 264–6). Is the experienced present durationless? As Dainton has pointed out, there is an argument from presentism to the durationlessness of the experienced present. He calls it the 'Augustinian' argument, though the implication is that this line of thought is suggested by Augustine's discussion rather than explicitly articulated in it. The argument is that if (as the presentist asserts) only the present is real, and (as Augustine has argued) the present is durationless, it follows that any experience, *qua* real event, is itself durationless (Dainton 2000: 120). However, as Dainton persuasively argues, the idea that experience consists of a series of literally instantaneous conscious states is deeply implausible:

Can we really distinguish, in introspection, an infinite number of distinct phases of a single short tone, or a perceived movement? Is there any intro-spective evidence that we can distinguish even a hundred? Physicists currently believe that intervals below the Planck duration of 10–43 seconds have no physical significance—is it likely that such intervals have any phenomenological significance? (Dainton 2000: 170)

He goes on to draw attention to experiments on temporal discrimination. Two stimuli are presented to subjects in very quick succession. When the temporal gap between them is sufficiently short, they are not perceived as occurring at distinct times. The point at which discrimination is no longer possible is known as the 'coincidence threshold'. For flashes of light or other visual stimuli the threshold is around 20 milliseconds; for auditory stimuli it is 2–3 milliseconds. If our capacity for temporal discrimination within experience had no limits, however, we should not expect any (non-zero) coincidence threshold for external stimuli. Here is another illustration of the limits to our capacity for temporal discrimination: a bright dot moving with sufficient rapidity over a screen will not be perceived as a dot at all; rather, the whole screen will appear illuminated. The dot occupies successive positions which are nevertheless perceived as simultaneous.

The experienced present may not be durationless, but it is certainly very short indeed. It is important, therefore, to distinguish it from what is generally called *the specious present*. The term 'specious present' was first coined by the psychologist E. R. Clay, but it was made famous by William James. As James describes it in *The Principles of Psychology*:

We are constantly aware of a certain duration—the specious present—varying from a few seconds to probably not more than a minute, and this duration (with its content perceived as having one part earlier and another part later) is the original intuition of time. (James 1890: 603)

The surprisingly large range of intervals which specious presents are supposed to span suggests that James has in mind the period which we are able to keep at the forefront of our consciousness, rather than the period of time that we actually perceive. Admittedly, Clay's original articulation of the idea suggests otherwise:

All the notes of a bar of a song seem to the listener to be contained in the present. All the changes of place of a meteor seem to the beholder to be contained in the present. (Quoted in James 1890: 574)

This cannot be true literally. We could not experience an entire spoken sentence as present: it would just be a confusing jumble of sounds if we did. (Think of the parlour game in which a group of people, in the absence of whoever's turn it is, chooses a well-known phrase, and, on his or her return, calls out the words of the phrase simultaneously, each member calling out a different word. The game, of course, is to guess the phrase, and as anyone who has played it will know, it is remarkably difficult to do so.) On the other hand, we are able to hold whole sentences in our minds, it seems, for otherwise we would not understand what was being said. Imagine if, by the time we heard the last few words of a sentence, we had to make a conscious effort to recall what the first words of it were. The chances of being able to grasp the meaning of the whole sentence, or at least to register it as quickly as we do in order to be able to attend to the next sentence, would be very slim. Nevertheless, we are clearly calling on our short-term memory here. By the time we hear the last word, the first is no longer being perceived, and can only be recalled from memory. But this process is so immediate and unconscious that we are not aware that this is what we are doing.

James's interpretation of Clay's idea involves awareness not only of the past, but of the future also. In an earlier passage, he provided the following characterization of what he called 'the practically cognised present':

> The practically cognised present is no knife-edge, but a saddle-back, with a certain breadth of its own on which we sit perched, and from which we look in two directions into time. The unit of composition of our perception of time is a duration, with a bow and a stern, as it were—a rearward and a forward end. It is only as parts of this *duration-block* that the relation of *succession* of one to the other is perceived. We do not first feel one end and then feel the other after it, and from the perception of the succession infer an interval of time between, but we seem to feel the interval of time as a whole, with its two ends embedded in it. (James 1890: 574)

This notion of the specious present, then, is very different from the idea we introduced earlier of the perceived present. James says that we distinguish earlier and later parts in the specious present. But we could hardly both perceive two events as present *and* perceive one as occurring before the other. What we experience as present is not divisible in this way. No part of what is perceived as present is simultaneously perceived as past and future. One way of understanding James's specious present, then, is to take it as involving a mixture of perception, memory, and anticipation. The term 'cognised present' certainly suggests more than raw perception.

Sean Kelly (2005), however, interprets James's doctrine of the specious present as a thesis specifically about perception, and contrasts it with the memory- (or retention-) based accounts of temporal experience offered by Kant and Husserl. According to Kelly, the doctrine of the specious present is the view that we directly perceive duration. Ultimately, he finds this notion unintelligible because it is quite unclear how we can *directly* perceive something that has ceased to be (in the case of the past) or has not yet occurred (in the case of the future), and it is also unclear how we can experience duration itself, as something separate from the event we are perceiving. Further, even if we could make sense of the doctrine, it does not help us understand the phenomenon it is often invoked to explain: namely, perception of change—an objection we will look at in §5.4.

Kelly's objections to the direct perception of duration seem to me to be entirely just, but I am not entirely persuaded that this is quite what James had in mind. For the view that something, whether duration or a tree, is directly perceived, without any intermediary like a sense-datum, is a philosophical thesis about perception, rather than a phenomenological thesis supported by introspection or psychological experiment. And a phenomenological thesis is what is suggested by James's remarks. He talks of what we 'seem to feel', not what we are in direct contact with. As to whether James take the specious present to be a purely perceptual phenomenon or one involving memory, his discussion is unfortunately rather ambiguous. On the one hand, he says: that

how long we may conceive a space of time to be, the objective amount of it which is directly perceived at any one moment by us can never exceed the scope of our 'primary memory' at the moment in question. (James 1890: 600)

(Here, admittedly, he *does* explicitly talk of directly perceiving an amount of time, which supports Kelly's interpretation; but the invocation of memory confuses things. Why would directly perceiving something, in the sense of the direct theory of perception, involve memory?) On the other, he wants to suggest a quasi-perceptual mechanism, akin to the persistence of after-images, as the basis for the extension of the specious present into the past (James 1890: 597–8). I shall have more to say about this later.

Although we have distinguished the experienced present from James's specious present, it seems that the experienced present is 'specious', in the sense of not presenting the objective present, in two ways: first, it presents the past, and second, it presents an interval, not an instant.

Can we avoid this conclusion? We might be able to preserve the idea that we perceive the objective present by adopting an account of perception according to which the immediate object of perception is a collection of sense-data. But if there is *any* causal gap between the sense-datum and the perceptual state, then there will also be a temporal gap (unless, what is arguably demonstrably false, cause and effect can be simultaneous), and the object of perception will still be a past object. But if there is no causal gap, then sense-datum and perceptual state are one and the same: the perception has *itself* as its object. Perhaps this is not in itself problematic, but a reasonable interpretation of the sense-datum theory. However, it does nothing to make more plausible the idea that the experienced present is durationless, for if we do not even mediately perceive durationless external events, how are we able to immediately perceive durationless sense-data?

There is a further reason to suppose that the sense-datum theory is not going to help the A-theorist here. Sense-data are supposed to be fallible representations of the world. They are introduced precisely to explain the fact that world and perception sometimes diverge. So, granted that the sense-data represent themselves as present, it does not follow that they actually *are* present, in the sense required by the A-theory, any more than the fact that they represent things as green means that the sense-data themselves are green. (Of course, one could construct a sense-datum theory according to which the sense-data exhibited every property they represented. But who would believe such a theory?)

The sense-datum theorist might still insist that it is only because they *are* present that the sense-data are perceived as such. Past and future sense-data are not perceived. But this is just the general point that our experience is temporally limited, to be discussed in the next section.

What if we now look at the presence of experience from the perspective of the B-theory? 'The present' no longer denotes an ontologically privileged, though durationless, moment, dividing past from future. It is now nothing more than a context-dependent term or concept, picking out whatever time is the temporal location of the utterance or thought in which it occurs. So if I think 'The clock is now striking', the 'now' is just the time I have that thought. Note that there is no temptation on this view to confine 'the present' to a moment of no duration. The time in question will have vague boundaries, to the extent that the thought has vague boundaries, but it will, like the thought itself, have some non-zero duration. So if I say 'This experience is of the present', or 'I am having this experience now', this amounts to no more than the triviality that I am having the experience at the time I am having it.

So if we interpret 'We perceive only the present' in A-theoretic terms, it is non-trivial, but false, and if we interpret it in B-theoretic terms, it is true, but trivial. Either way, it is not a datum from which we can justifiably infer significant metaphysical consequences.

5.3 INTERSUBJECTIVE AGREEMENT AND THE
TEMPORAL LIMITS OF PERCEPTION

That our perceptual experience has temporal limits is undeniable: what we are attending to in perception—the arrow hitting the target, the flash of lightning, the opening notes of a symphony—are events that occur at a particular time, not a significantly extended series of events. To contemplate such a series requires the resources of memory and imagination (which is not to say that these play no role in what we call 'perception'). In the appropriately narrow sense of 'perceive', we perceive neither the (distant) past nor the future. Moreover, observers in the same spatial location tend, on the basis of their perceptions, to agree on what is going on now. The A-theorist has an explanation for all this: it is only when events are present that we can perceive them, and since we share the same present, what is happening now *is* the same for all observers, and so, by and large, what *appears* to be happening now is also the same. But if it has already been conceded (the sense-datum theory aside) that only past events are perceived, then this explanation is undermined.

Perhaps, then, a more appropriate A-theoretic explanation is this: 'Experience *does* lag behind external events, but when an event becomes present, that sets in motion a causal process that eventuates in the perception of that event. With the passage of time, however, the original event is replaced by another, which sets another process in motion, leading to another perception which thus displaces the first. Once an event ceases to be present, it loses the power to set anything in motion, and that is why we do not perceive the more distant past. As for intersubjective agreement, since we all share the same present, roughly the same information will impinge on us from the same events. Contrast this with the B-theory, according to which all events, whenever they occur in time, are equally real, and so, one might suppose, equally capable of giving rise to perceptions, however temporally distant they are from the perceiver. Further, since observers are not restricted to the objective present—there being, for the B-theorist, no such thing—the B-theory gives us no reason to expect that all observers will agree.'

(This, incidentally, would provide one response to Simon Prosser's (2007) argument that since physics makes no appeal to the passage of time, and causation is a purely physical relation, the putative passage of time is purely phenomenological. In the above account, the presentness of an event is a necessary condition of its causal efficacy.)

This is a significant challenge to the B-theorist, whose wisest strategy is to concede both that we have found a genuine feature of our experience and that the A-theorist has a genuine explanation of it, but then to go on to provide a B-theoretic explanation of equal power.

Let us start, then, with the fact that we do not perceive the future. The explanation for this, the B-theorist can say, is purely causal: causes always precede, never succeed, their effects in time. And perception is a causal process: the object of perception is the cause, the perceptual state the effect. To 'perceive the future', in B-theoretic terms, would be to perceive events that are later than the perceptual state. But this would involve backwards causation, which we assume to be impossible. Therefore, perception is only of the past—i.e. of events that occur earlier than the perceptions they cause.

(Admittedly, we do not *hallucinate* the future either, and this cannot be explained in similar causal terms, since the object of a hallucination, being non-existent, is not in a position to cause anything. The explanation, in this case, is purely phenomenological. Hallucination is illusory perception, and what we perceive—or seem to perceive—we perceive as present. Consequently, the object of a hallucination does not present itself as future.)

As the A-theorist justly points out, however, all past (earlier) events are equally real for the B-theorist. Why, then, at any one time, do we perceive only a limited part of the past? No part of the past is especially privileged, we should note. Any past event, however distant, is a potential object of perception for some suitably situated observer. Which part of the past is perceived will depend on the observer's spatial distance from the event in question. The further away the event in space, the greater the temporal gap between event and perception. This is so because there is no unmediated action at a spatial and temporal distance. Past events cannot directly cause perceptual states in us: they must do so via a series of intermediate causes. Only when information from those events has reached us across whatever spatial distance separates us from the event in question, will we begin to perceive it. So our spatial position is part of the explanation for the temporal limits of our perception. If an event occurs four light-years away from us, we will not perceive it for four years.

Why, then, once information has reached us from an event, do we not continue to perceive it? Why should we not, at any moment, perceive all the events that have ever impinged on us? There is nothing incoherent in this suggestion. Beings who perceived the world like this are a logical possibility. But they would be at a massive biological disadvantage. Responding appropriately to its environment requires an organism to register what is going on in its more or less immediate vicinity at that time. That, and the limited storage capacity of information-processing systems, requires that, once an item of information is registered, it must be moved quickly into the memory to make room for incoming information. An organism that failed to do this would, when once faced with prey, predator, or mate, find itself continuously perceiving that object, even after it moved away from the vicinity. Such an organism would therefore rarely respond in a timely fashion to the presence of significant objects in its environment, and would very quickly indeed fall victim to the forces of natural selection.

It remains to explain intersubjective agreement. If we do not share an objective present, why do we tend to agree on what is happening now? The question to ask here is what population is picked out by 'we'. It is quite true that, from the B-theoretic perspective, all observers, wherever located in time, are equally real. But the 'we' of 'we tend to agree' does not pick out *all* observers: it picks out only a certain subset, namely the individuals with whom the speaker is, or could be, in communication. And we are in direct communication (i.e. in a way that is not dependent on technological intervention of any kind) only with those people with whom we share a temporal perspective. If I speak to you at t, then you hear me almost (but not quite) at t, since sound travels very fast. And anything in the reasonably close vicinity occurring at t will impinge on us almost (but not quite) at t. So, within a very short interval indeed, we will be able to sense what is happening at t and communicate it to each other. And, given that the speeds of information exchange both from object to observer and from observer to observer tend to be much faster than the rate at which most observable things change, we will generally be right in our perceptual judgements and utterances (in normal circumstances). Were the speeds of light, sound, and information processing and exchange to be much slower in relation to the rate at which things change, we would not experience this intersubjective agreement, and so perhaps lack the sense of sharing the same moment. (The explanation offered here is due to Butterfield (1984).) It seems, then, that our common temporal perspective may be due to entirely contingent features of the physical world and our neurophysiological constitution rather than some necessary truth about time.

In sum, by appealing to a variety of principles—the temporal asymmetry of causation, the absence of unmediated action at a spatial or temporal distance, the limitations of information-processing systems, the forces of natural selection, and the rates of information exchange—the B-theorist can explain the temporally limited nature of perception. Of course, this explanation is available to the A-theorist too, but appeal to it would make the A-series simply *de trop*.

5.4 A PHENOMENOLOGICAL PARADOX?

'We perceive change and motion.' This appears metaphysically signifi-cant, because it is what 'the immediate awareness of the passage of time' amounts to. We are indirectly aware of the passage of time when we reflect on our memories, which present the world as it was, and so a contrast with how things are now. But much more immediate than this is *seeing* the second hand move around the clock, or *hearing* a succession of notes in a piece of music, or *feeling* a raindrop run down your neck. There is nothing inferential, it seems, about the perception of change and motion: it is simply given in experience. As Broad puts it,

> to see a second-hand moving is quite a different thing from 'seeing' that an hour-hand has moved. In the one case we are concerned with something that happens within a single sensible field; in the other we are concerned with a comparison between the contents of two different sensible fields. (Broad 1923: 351)

What is interesting about this phenomenological datum is that it is in tension with another datum in the list presented in §5.1: that what we experience, we experience *as present*. If we have a single experience of two items as being present, then, surely, we experience them as *simultaneous*. Suppose we are aware of A as preceding B, and of B as present. Can we be aware of A as anything other than past? Of course we can have successive experience of items, each of which, in turn, we see as present. But no single experience presents all these items as present.

But now consider what happens in motion perception: we see an object occupying *successive* positions. We must see these as non-sim-ultaneous, for otherwise we would just see a blur. (Of course, when motion, especially oscillatory motion, is sufficient fast, we do see the object as a blur. But standardly, the moving object is in sharp focus.) It follows from this that we do not see these successive positions as present. But if the perception of motion just *is* the perception of successive positions, then it seems that we do not, after all, perceive motion as present, which is to say that we do not perceive it at all. Yet how can we deny such an obvious feature of our experience? Just *look* at the birds flying overhead, the bus moving off down the lane, the leaves tumbling down from the trees. This motion is, it seems, phenomenologically present.

So what do we say in order to avoid this phenomenological paradox? That not everything we perceive we perceive as present, or that in perceiving motion we are not perceiving successive positions? Neither of these options seems at all comfortable. Perhaps, then, we should just accept that the phenomenology of motion *is* paradoxical, that we really do experience contradictory things. We could say this, perhaps. After all, the injunction to treat experience as consistent is a good deal less stern than the injunction to treat the world as consistent. But if the data here are indeed contradictory, then their support for the A-theory is correspondingly compromised. They cannot *both* be represented as intimating how things are. If we really do perceive successive positions in a single perception, then that perception cannot be of the present. If, on the other hand, perception really is limited to the present, then it is incorrect to talk of perceiving motion, change, and the passage of time (even if this is what we seem to be perceiving).

The suggestion that there is a phenomenological paradox here may induce some discomfort, however. Ordinary motion perception does not seem at all like those cases of perceptual illusion where we are aware of some inconsistency. Take a case where we clearly are presented with phenomenologically inconsistent data: the waterfall illusion. If, while keeping one's eyes as still as possible, one looks for a long time at continuous movement in a particular direction—as when, for instance, we watch a waterfall, or railway tracks flying past when we are looking out of the carriage—and then turns one's gaze to a stationary object, it will seem for a while to move in the opposite direction. A similar effect is produced by looking at a rotating spiral. While it is rotating, it will appear to expand (or shrink, depending on whether it is an anti-clockwise or clockwise spiral, and whether it is being spun clockwise or anti-clockwise). If it appears to be expanding, then, once the spiral stops, it will appear to shrink (and vice versa). As Richard

Gregory points out, the experience is paradoxical: the spiral seems to remain the same size *and* change size at the same time. In the case of linear movement, the stationary object seems both to move and to stay in the same position. How is this? Gregory's suggestion is that two neural mechanisms are involved here, one for detecting motion and the other for detecting change of position. The motion-registering system is somehow disturbed by the previous exposure to movement, and continues to indicate movement when all external motion has stopped. The change of position-detecting system is not disturbed in this way. The two systems then give contradictory information when presented with a stationary object (Gregory 1966: 104–9). This explanation of the effect may undermine the suggestion (Crane 1988) that the illusion obliges us to treat the content of the perception as non-conceptual (Mellor 1988).

Just as an aside: we may initially be puzzled by Gregory's explanation. After all, what is motion except change of position? So how can a system register motion without at the same time registering change of position? It is true that the cause may be the same in both cases, but one system may be more primitive than the other. Suppose motion to be registered simply by a change in retinal stimulation. A more sophisticated system may register the relative position of an object and store it in the short-term memory for comparison with later perceptions of its relative position. Any discrepancy would result in the perception of change of position. It is possible, then, that the more primitive system would register motion, while the more sophisticated system would report no change in relative position.

Ordinary motion perception, however, is, as we have said, not obviously paradoxical in this way. Yet there does seem to be a contradiction in how we report it: we see the motion as present, *and* we see successive positions, which therefore cannot all be seen as present. Perhaps we can apply Gregory's explanation to this case too: what we have here are two neural mechanisms in play. One system registers what we might call 'pure' motion, i.e. gives rise to the impression of motion without any associated sense of change of relative position. It is this system that is responsible for the sense of perceiving motion as happening *now*. Another system, the one that employs short-term memory, takes a series of snapshots of an object's relative position and compares them. That system gives rise to the sense of change of relative position, but it cannot unproblematically be said to give rise to the sense of change of relative position happening *now*.

That would be a consistent (if speculative) explanation of what is going on in motion perception, but it would need to be generalizable if it is to resolve the paradox. For we are also aware, in an apparently non-inferential way, of change in pitch when we are listening to a piece of music, or of the passage of thoughts when we are doing nothing more than daydreaming. Are there two mechanisms underlying change detection in these cases, too? Let us consider two notes heard successively, say a C followed by an E. How are these presented in experience so as to give rise to a perceptual impression of a musical phrase? Let '[Present (x)]' stand for a mental representation of x as present, and, '[Past (x)]' for a mental representation of x as past. Then the following schemata seem to exhaust the possibilities in respect of this case:

(1) [Present (C)] simultaneously with [Present (E)]
(2) [Present (C)] followed by [Present (E)]
(3) [Past (C)] simultaneously with [Present (E)]

None of these seems to give us what we are looking for. (1) is what we get when we perceive the two notes together, and not as successive. (2) is simply the succession of perceptions, not the perception of succession. And (3) gives us, not a perception, but a perception presented together with a memory—the kind of case that Broad would describe as the awareness *that* E had been preceded by C.

Does the doctrine of the specious present help us here? The doctrine does not introduce a fourth schema, so much as point to two ways of interpreting schema (3). Here is James again:

With the feeling of the present things there must at all times mingle the fading echo of all those other things which the previous few seconds have supplied. Or, to state it in neural terms, there is at every moment a cumulation of brain-processes overlapping each other, of which the fainter ones are the dying phases of processes which but shortly previous were active in a maximal degree. (James 1890: 597–8)

So, instead of a memory of C, [Past (C)] could be a *fading perception* of C, and it is the diminished intensity of the perception that conveys to us the sense that C is past, while keeping it in the compass of our sensory experience. In Broad's version of the doctrine, we have a series of acts of awareness, each of whose contents overlap (Broad 1923: 348 ff.; see also Broad 1938 for a fuller account).

One difficulty with this account, as Sean Kelly has pointed out (Kelly 2005: 222), is that it fails to account for perception of change and motion over a period longer than that of the specious present. What is it that links the different specious presents together, providing a sense of continuous motion over a period? It cannot be a specious present itself, since no specious present has other specious presents as parts. Therefore there must be some other mechanism at play. Dainton, however, suggests that it is precisely the overlap between different specious presents that links them (Dainton 2001: 102–5). The continuity of experience is explained by the very large number of such overlapping experiences, even in very brief perceptions, say of the two notes. But the additional element that Dainton requires is what he calls 'an inherent directional dynamism': each experience has an in-built direction (p. 105).

The doctrine of the specious present, however, does not allow us retain the datum that what we perceive we perceive as present, and this is what, ultimately, makes it so implausible. The past events are still perceived, but as fading images, and this loss of vivacity, as Hume would put it, conveys the feeling of pastness. But after-images do not convey feelings of pastness at all: what we perceive are the images themselves, and these are experienced as present. So if the doctrine of the specious present were correct, we would, on hearing the second note, still be hearing, very faintly, the first. This is not only false, but does not explain the sense of succession.

Broad and James interpret the doctrine in different ways. Broad, unlike James, does not permit awareness of the future to intrude into his specious present, which is, in addition, much shorter than James's maximum of a minute. And James's discussion is, as I have said, ambiguous. The passage quoted above certainly suggests a purely perceptual interpretation, but elsewhere memory is brought into the picture. And this takes us to the second, and most natural, interpretation of schema (3): the perception of E is accompanied by the memory of C. Because of the proximity of the perceptions, however, the experience of succession is not consciously inferential. The thinking behind the specious present, I suggest, is that to perceive at any one time the succession of notes, I have to perceive both notes at that time. But this is mistaken. What I want to suggest is that the conjunction of the very recent memory of C with the perception of E gives rise to an experience of 'pure succession'. That there should be such a sense is not at all implausible when we reflect on the existence of the primitive mechanism that Gregory appealed to for pure motion detection, which is distinct from the mechanism for detection of change of position. I am not suggesting that a similar mechanism exists for sounds (though it may). I am simply putting forward the phenomenological thesis that we perceive succession of notes in a way that can be distinguished from perceiving C being followed by E. The latter, though not the former, involves a relation, and since such a relation could not be perceived independently of the relata, the suggestion that we perceive E's following C inclines us to suppose that C and E have somehow to be perceived together. What gives rise to the experience of pure succession, in contrast, is the conjunction of the perception of E with the very recent memory of C.

This, I suggest, is how we might resolve the phenomenological paradox. It is an explanation that favours neither side of the metaphysical debate. It is consistent with both the A-theory and the B-theory. But its very neutrality shows that there is nothing in these data to incline us towards the A-theory.

Let me end this section on a cautionary note: we should be careful not to explain *too much*. The description of the phenomenology should not be too clear-cut. For the truth of the matter may be that experience does indeed give us the impression of change and succession, but no very clear sense that this belongs to a particular moment. We are aware of it, and that is all. Describing succession as phenomenologically present does not tie it down to a moment. Allied to this is the thought, urged by Dennett (1991), that there is no definite moment at which a piece of information becomes conscious. Here is a little experiment which helps to illustrate this. Read the following sentence out aloud to a friend (but do not allow him or her to see the written text beforehand): 'Rapid righting with his uninjured hand saved from loss the contents of the capsized canoe.' (The example comes from the psychologist Karl Lashley, and is cited in Gombrich 1964: 301. It is used to make a somewhat different point from the one made here.) The first part of this sentence, as spoken, is compatible with 'writing' as the second word. But that interpretation does not make sense of the second part. Now ask your friend: 'When you first heard "righting", did you know what was meant? Was there an identifiable point at which you realized that "righting" (i.e. correcting) was the word, and not "writing" (i.e. inscribing)?' The chances are that your friend will have some difficulty answering these questions. Is this due to the limitations of memory, or that there was no fact of matter when a particular perceptual experience was had?

5.5 MOTION, PASSAGE, AND PROJECTION

An assumption implicit in the argument from experience is that there is a direct correspondence between the perception of change and the objective passage of time, or its consequences. Change and motion are out there, and we just register them. But as perceivers are we really so passive? Is there not an element of the mind's construction in what we experience? According to a very influential theory of motion perception, proposed by Herman von Helmholtz, what we see when we see motion is in part due to the mind's telling us what we see. We see motion when we keep our eyes still, and a moving object projects a shifting image across our retinas. But we also see motion when we track the moving object with our eyes, so that the retinal image stays constant. The systems responsible are known as the image/retina system and the eye/head system. These systems have to co-operate all the time. For, as we take in a static environment by sweeping our eyes over it, as we often do, our retinal images are constantly shifting. But the world outside does not appear to move. Why? It seems that information from the two systems can cancel each other out: the brain registers the changing image, but attributes the shift simply to the fact that the eyes are moving, so we do not register motion. This involves some interpretation on the part of the brain, but it only works if the movement is voluntary. If the eyeballs are moved mechanically, e.g. by pressing gently with the finger, then motion is registered. This suggests that the process of comparing information from the eye/head and image/retina systems does not depend on information passively received from the eye muscles, but rather on information from the part of the brain responsible for controlling eye movement. This explanation is further confirmed by experiments in which the eyeballs are prevented from moving. When the subject attempts to move them, the world appears to shift. Here, then, is a case of perceived, or apparent motion, in the absence of both eye movements and retinal image changes. The brain is simply telling itself that there must be motion out there, and the visual cortex dutifully registers it. (For a fuller account, see Gregory 1966: 90–9.)

This datum of experience, that we perceive motion, is, then, no mere passive reception of an objective phenomenon, but one that arises, in part, from the active interpretation of the mind. It is, at least in some cases, a *projection*.

The final case I want to discuss in this chapter is known as the 'flash-lag' phenomenon. Subjects are presented with a small dot moving across a screen. At some point during the dot's transit, another dot appears very briefly directly above or below the other dot: this is the 'flash'. A significant number of subjects report that they saw the moving dot *ahead* of the flashed dot, in the direction of movement. The flash appears to lag behind (McKay 1958). But it does not really do so, so why the illusion that it does?

Here is one, very plausible, explanation of the phenomenon. As we noted above, it takes time for the brain to register what is going on in the outside world. When information is registered about changing objects, and especially about rapidly moving objects, it is already slightly out of date. So we have evolved to compensate for this lag. The brain makes an adjustment to the information it has received about the position of a moving object, and makes a projection based on information concerning the object's velocity and direction. When we see the object, we see it, not in the position it was in when light from it hit the retina, but in the position the brain estimates that the object must be in by the time that information is registered. What we see, once more, is a projection, based on a prediction. Now this mechanism is in play when subjects see the dot moving across the screen. They have registered that it is moving, and so make the necessary compensation by slightly advancing the position of the dot. But the perception of the flash is not subject to the same compensation, since there was no prior information about its position. So its position is not advanced, and it appears to lag behind the other dot (Nijhawan 1994).

Contrasted with this 'predictive' model is the 'postdictive' model, according to which what is perceived depends on what happens imme-diately *after* the flash. Some experiments suggest that, if the moving dot changes trajectory after the flash, it is the post-flash, not pre-flash, trajectory that determines the resulting visual effect. According to the postdictive model, this is because the various data are integrated by the brain and then backdated to the moment of the flash: the brain, as we might put it, decides what it *has* seen. (For a report of the relevant exper-iments, and the postdictive interpretation, see Eagleman and Sejnowski 2000.) As with the predictive model, however, the suggestion is that the brain is imposing an interpretation on the data.

The moral of these various findings, and their (admittedly debat-able) psychological or neurological interpretation, is not that motion perception is essentially illusory, but that there is at least a component of motion perception that is constructed, or projected, by the mind. Further, that there are cases where the mind projects motion where there is none. So when the A-theorist points to the fact that we just perceive motion in the world, as support for the notion that time passes, we should note that this is not always a datum that is simply forced on the passive mind by external circumstances, but is actively constructed, at least in part. It does not follow that motion is always projected. Normally, we would suppose that the best explanation of the experience as of motion is that we are genuinely presented with motion. But it is salutary to reflect that the perception of motion is more complex than it might appear, and this should encourage anyone to exercise some caution in framing arguments from experience in favour of the A-theory.

The considerations of this chapter also pave the way for a rather more robust conclusion. If the argument from experience is that the A-theory provides the best explanation of temporal phenomenology, then this has certainly not been confirmed by our exploration of the data: explanations are available which do not appeal to A-series position. And this provides some motivation for a B-theoretic outlook: A-theoretic properties are not in the world, but are *projected* on to the world in response to certain features of our experience. This would be closely analogous to projectivist views of secondary qualities: the world itself is not coloured, but certain properties of objects induce in us sensations which cause us to ascribe colours to them (see e.g. Boghossian and Velleman 1989). Gale, however, is sceptical of such an analogy:

As far as I can make out, no defender of the mind-dependency thesis has advanced anything which is even remotely analogous to the . . . causal theory of perception [of secondary qualities]. That they have not done so is not a result of laziness or an oversight on their part, but rather is due to a basic difference between *A*-determinations and secondary qualities which makes it impossible to advance an analogous causal theory of the perception of becoming. The crucial difference is that *A*-determinations are not sensible properties, whereas secondary qualities are. (Gale 1968: 228)

Even if pastness, presentness, and futurity are not sensible properties in the way that colours are, however, the argument from experience (a version of which Gale himself provides) is based on the conviction that it is features of sensory experience, including motion perception, that convey to us most powerfully the sense of the passage of time, and that this is best explained by the objective passage of time. If we can provide an alternative explanation of this experience, one which appeals crucially to causal factors, as we have done in this chapter, then this does provide an argument in favour of projectivism about passage. And if the analogy with colours is not perfect, we might compare projectivism about passage to projectivism about moral values. Moral values are not sensible properties in the way that colours are. But projectivists can appeal to causal considerations, such as the emotional effect that certain natural properties of events have on us, in defence of the view that the world does not contain objective, mind-independent, moral values (Blackburn 1984). Gale's scepticism, I suggest, is not justified.

6
The Wider View: Precedence and Duration

DR GÖRTLER: We have to change the focus of attention, which we
have trained ourselves to concentrate on the present. My problem
was to drift away from the present—as we do in dreams—and
yet be attentive, noting everything—
FARRANT [with savage intensity]: Yes, yes, but how did you do it?
By going without food, I suppose?

J. B. Priestley, *I Have Been Here Before*

6.1 THE PUZZLE OF TIME PERCEPTION

I am looking at the clock on the mantelpiece, and note that both hands
are pointing to twelve. Here, surely, is a straightforward case of veridical
perception. There is the clock, and I am looking at it in near-ideal
conditions. Without question, I see the clock, and the position of the
hands, and at least a case can be made that I do so in an apparently
unmediated way. The direct theory of perception is, if applicable to any
case, applicable to this one.

But now the clock strikes noon, and I perceive a host of other things:
not merely a series of sounds, but one chime as *following on* from
another, the *interval* between chimes, and that interval as remaining *the
same* in each pair of chimes. All these are instances of time perception,
in that the content of the perceptions seems irreducibly temporal. But
perceiving time in this way, and perceiving the clock, seem very different
kinds of experience. Even if we confine our attention just to veridical
time perception, when our senses do not deceive us as to the order or
duration of events, the direct realist approach seems much less tempting
than in the case of ordinary perception of objects. In fact, it is not clear
that direct realism about time perception is even an option. Why is this?

According to the title of an address given by the psychologist J. J. Gibson to a meeting of the International Society for the Study of Time in 1973, 'Events are perceivable but time is not' (quoted by Pöppel 1978: 713). In so far as we think of time as something independent of the events within it, as an unseen, featureless medium, we could hardly fault this description. But whether or not time is independent of its contents, events themselves have temporal features: they occur in a certain order, and they last for a certain amount of time. And if we were completely insensitive to these features, then it would be a mystery how we come to be aware of motion, interpret Morse code, or appreciate music, to give just a few examples. However, it is one thing to form beliefs about order and duration on the basis of our perceptions, and quite another actually to *perceive* order and duration. If the meaning of Gibson's dictum is that, although we may come to be aware of the temporal features of events, those features are never themselves the direct objects of our perceptual states, then a number of considerations tell in its favour.

First, there is a disanalogy between awareness of spatial features, such as size or shape, and awareness of temporal ones. There is a relatively straightforward story to be told about the way in which the shape of, say, an apple, can be an object of perception when the apple is (although the details of the story may be complicated): the shape of the apple modifies the distribution and properties of light rays reflected from its surface and which then hit the retina; it also modifies both the distribution of pressure exerted by the apple on our touch receptors and the kinaesthetic input as we close our hands around it. With time, in contrast, it does not make sense to suppose that, when a given event such as the ringing of the telephone is perceived, the duration of that event, or its occurring after some other event, somehow modifies the input, allowing us just to *hear* its duration and position vis-à-vis other events. We are only aware of how long an event lasted when it has receded into our phenomenal past—when, in other words, the event has ceased to be an object of perception.

Second, as we noted in the previous chapter, perception has temporal limits, and if we draw these very tightly, certain things cannot be objects of (at least direct) perception. If what we perceive, we perceive as present, then it seems that we do not perceive either temporal order or duration. For to perceive these is to recognize earlier and later parts in what we perceive, and, as we also noted in the previous chapter, we cannot recognize earlier and later parts in what we see as present. So, if we do perceive order and duration, it seems that we do so in a way that is not analogous to perceiving something as happening or existing *now*.

Third, whereas it is possible to be aware just of the colour of an object—I can focus on its redness, say, to the exclusion of all else—it is not possible to be aware just of the relation of temporal subsequence when it obtains between two events. To see one event as following on from another, I have to be aware of the events themselves. The relation by itself cannot be an object of perception as pure colour can. Similarly with duration: I cannot be aware *just* of the duration of an event, independently of my awareness of the event itself. Somehow, the awareness of order and duration emerges from a perception of the events that exhibit them.

I did suggest, in the previous chapter, that we have a sensation of pure succession. But I distinguished this from precedence, which is essentially relational, and therefore not detachable in perception from the relata.

So, on the one hand, our perceptions inform us of the order and duration of events. Moreover, the process appears to be a non-inferential one: we can become aware of temporal features without having consciously to derive them from other information. But, on the other hand, the usual models for ordinary perception, of shape and colour for example, are not applicable to time. In particular, order and duration are not in any straightforward sense objects of perceptual states. As Woodrow (1951: 1235) puts it, 'time is not a thing that, like an apple, may be perceived.' What, then, are the mechanisms underlying our perceptual awareness of time order and duration? This is the *psychological puzzle* of time perception. Here is one psychologist's perspective on the problem:

There is no process in the external world which directly gives rise to time experience, nor is there anything immediately discernible outside ourselves which can apprehend any special 'time stimuli'. It is therefore not too surprising that psychological research on time as a dimension of consciousness has been so diverse, so incoherent, and so easily forgotten. (Ornstein 1972: 96)

(It is only fair to point out, in relation to the second part of this quotation, that Ornstein was writing more than thirty years ago, since when there has been a considerable amount of research on time perception. And even at the time, it might have been considered a less than generous remark. (See e.g. Pöppel 1978 and Friedman 1990 for references.) The second part of the first sentence in this quotation is rather puzzling. I am inclined to think that when Ornstein says 'outside ourselves' he means '*within* ourselves' (i.e. in contrast to the external world), and that his point is that there is no obvious sense-organ for time. The sentence, however, is not corrected in later editions.)

Of course, a psychological problem requires a psychological answer, not a philosophical one. But there is a philosophical dimension to the issue, one concerning the epistemological status of our perceptual awareness of the temporal features of the environment. If we can take it as a datum that we can form, on the basis of our perceptions (by some mechanism yet to be determined), beliefs concerning both time order and duration, we can also take it as a datum that these beliefs are at least capable of being, and generally are, a form of perceptual knowledge—although of course we often make mistakes. Now, according to the influential and plausible theory of perceptual knowledge discussed in Chapter 2, the truth-makers of those beliefs should occur in the causal history behind our acquisition of them. This is what we called the 'Causal Truth-Maker Principle' (CTMP). But one moral of the disanalogies we were drawing attention to above between ordinary perception and time perception is that *time order and duration are not causes of our perceptual states*. And if they are not the causes of our perceptual states, then it is hard to see how they could be the causes of beliefs that arise from those perceptual states. This suggestion, that order and duration do not play a causal role, will be confirmed shortly when we come to look at models of time perception. It is a problem, then, to explain, compatibly with the causal theory of perceptual knowledge, how time perception can lead to knowledge of the temporal features of the environment. This is the *epistemological puzzle* of time perception. Solving this puzzle is the aim of this chapter.

Making things a little more explicit, the puzzle arises from the following conjunction of plausible and defensible propositions:

The Causal Truth-Maker Principle: Perceptual beliefs that qualify for the title 'knowledge' are caused by their truth-makers.

The acausality of order and duration: The objective order and duration of events are not the causes of our perceptual beliefs concerning order and duration.

Together, these two appear to imply that, even if there is such a thing as time awareness, it cannot count as knowledge, since the truth-makers of our beliefs in these cases play no role in the causal history of those beliefs. (To derive this result formally, we have of course to add what may seem the rather obvious proposition that actual order and duration, or facts concerning them, are the truth-makers of perceptual beliefs about order and duration.)

I have already provided an initial defence of the CTMP in Chapter 2, and it looks, if not invulnerable, at least something we would be reluctant to give up. It seems only sensible, then, to have a critical look at the second proposition.

6.2 THE ACAUSALITY OF ORDER AND DURATION

Even if order and duration are not *objects* of perceptual states, it does not immediately follow that they are not *causes* of those states. There are any number of hidden causes of our perceptual states—to do, for example, with the underlying psychological or physiological mechanisms of perception, which are never the objects of those states. But we are concerned with external features of the world, and it is hard to see how they can causally impinge on us if we never perceive them. However, there are other reasons for thinking that order and duration may not be the sort of features that are capable of entering into causal relations. I offer here, not a knock-down argument to the effect that they cannot possibly affect things causally, but rather some awkward questions for anyone who thinks that they do.

In a well-known discussion of the existence of universals in *The Problems of Philosophy*, Russell makes the following observation:

Consider such a proposition as 'Edinburgh is north of London'. Here we have a relation between two places, and it seems plain that the relation subsists independently of our knowledge of it. . . .

. . . however . . . the relation 'north of' does not seem to exist in the same sense in which Edinburgh and London exist. If we ask 'Where and when does this relation exist?' the answer must be 'Nowhere and nowhen'. There is no place or time where we can find the relation 'north of'. . . . Now everything that can be apprehended by the senses or by introspection exists at some particular time. Hence the relation 'north of' is radically different from such things. (Russell 1912: 56)

If this correctly characterizes 'north of', it also characterizes 'later than'. And, since duration is also a relational property, Russell's observation would apply to that too. But surely *causes* are items that have spatial and temporal locations. Their having those locations is what helps to explain why they have their effects at the times and in the places they do. But if order and duration relations do not have locations, they are acausal.

However, as he makes clear, Russell takes the 'north of' relation to be a universal, and therefore something that any number of places can stand in to each other. What if, instead, we concern ourselves with the *trope*, the token of the type, *this very instance* of 'north of', relating Edinburgh and London? What is its location? We still face difficulties in arriving at anything other than an *ad hoc* answer to the question. There seems no good reason to locate the relation just in Edinburgh, or just in London. Is it, then, in both? But then tropes would be quite unlike other particulars, which cannot be in two places at once. Perhaps, then, it is located in the region *between* Edinburgh and London. But how do we define this region? Is it just the straight line connecting the two? Is it the whole of the United Kingdom? Should it not include the North Pole? Is it the surface of the Earth? Even if we are dealing with the trope rather than the universal, perhaps the best thing to say is that the relation is *nowhere*: it does not have a spatial location at all. The same worries arise for order and duration. There is a clap of thunder at 4 o'clock, and the rain starts pouring down at one minute past 4. These events are readily locatable. But what of the temporal relation between them? That relation, even if we treat it as a trope, is not readily locatable at 4, or one minute past, or any time in between. And what of the rainstorm's property of lasting 20 minutes? Where is that in time? Again, we may be tempted to say that these properties have no temporal location: they exist timelessly. But if they are timeless, they cannot be causes. If we insist that these relational properties do have a location, only that they are somehow spread out in a spatio-temporal region that includes the relata, and that is how those properties can be causes, we are still faced with the question: Why, if they are spread out in space and time, are they not capable of having effects in all parts of that region? Whether or not some 'earlier than' trope is the cause of my belief that the thunderclap preceded the rain, it cannot possibly cause my belief *before* I experience the rain. But then why not, if it exists before the rain does?

It does not help to point out that duration and precedence are consequences of events' temporal locations, for this relies on those locations themselves being temporally extended and standing in temporal relations to each other.

Some further worries, not to do with location: Are there in fact any cases where we cannot explain away the apparent causal efficacy of temporal properties? Take the case of the egg's being cooked because it has been boiled for 5 minutes. What is doing the causing here? The mere passage of time? No: the egg is cooked because of the individual events that took place during those 5 minutes. Or take a machine that produces one output, C, if the two kinds of input, A and B, occur in the order AB, and another output, D, if they occur in the order BA. Is the mere order of the inputs *itself* a cause? Or is the correct story rather something like the following: given an initial state of the machine, S_1, the effect of input A is to change the machine's state to S_2. And when the machine is in S_2, the effect of input B is the production of output C. But when the machine is in S_1, the effect of input B is to change the machine's state to S_3, and the effect then of input A is not to change the machine's state to S_2, but rather the production of output D. On this account, it is not the mere order of inputs that determines the output, but what state the machine is in when it receives a given input. Now, if we are to take seriously the suggestion that order and duration *per se* can be causes, then there have to be cases where the natural form of explanation that applies in these cases of the egg and the machine is just not available. And this, intuitively, seems odd.

The difficulties of locating spatial extension raise similar causal worries. Yet we do, intuitively, think that spatial extension plays a causal role: the length of the telegraph pole is, surely, causally responsible for the length of the shadow it casts. But perhaps the right thing to say is that the spatial extension is just a function of the different parts of the pole, each of which has its own distinctive causal effect. Is all causation ultimately perhaps a matter of interaction between point-like and instantaneous states?

These are just expressions of puzzlement, of course. But given that treating order and duration as causal is problematic in one way or another, it would be best not to give those properties a causal role in our account of how we acquire order and duration beliefs. So we come back to the epistemological puzzle. In trying to solve it, it would help if we had some insight into the psychological mechanisms of time perception. It should not be necessary to have at our disposal a detailed description of this undoubtedly complex process. Since the puzzle we face is one that is raised at a high level of abstraction, it should be enough to show that some plausible abstract description of how we acquire beliefs concerning order and duration through perception of events can be reconciled with epistemological requirements.

6.3 THE PSYCHOLOGICAL BASIS OF TIME PERCEPTION

When we were considering motion perception in the previous chapter, we faced the problem that, to the extent that perceiving motion requires discrimination between earlier and later positions of the moving object, we cannot see motion as present. And yet, in a sense, we do see motion as happening now. In attempting to solve this problem, we appealed to a mechanism involving short-term memory. Perhaps we should do the same with perception of order and duration. In the case of simple precedence, Hugh Mellor has suggested the following mechanism: stimulus *a* is perceived, and its occurrence registered in the short-term memory. Stimulus *b* is subsequently perceived, and the representation of *a* in the short-term memory causes *b* to be represented as occurring after *a*. Perception of precedence therefore is not simply a matter of one perception preceding another, but that earlier perception causally affecting the later one (Mellor 1981: 143–50; 1995: 138–40; 1998: 114–15). Perception of precedence, then, arises from the same mechanism that we suggested in the last chapter is responsible for the sensation of pure succession.

If we can allow simple introspection here as judge, this account seems very plausible. Perceptions do seem to be coloured by immediately preceding ones. However, there are some apparent difficulties with this as a general account of the awareness of precedence. The first is raised by instances of 'backward referral', where the perceived order of two stimuli, as reported by the subject, is the reverse of the order of the perceptions. One rather striking case is reported by Benjamin Libet (1981) as follows. Tingling sensations in the hand can be produced in two ways: by electrical stimulation of the skin of the hand itself, or of the appropriate part of the somatosensory cortex. The latter, in effect, mimics the later stages of the neural events associated with hand stimulation, and so, one might expect, should lead more quickly to the perceived sensation. In one experiment, the subject's left cortex was stimulated very shortly before the left hand was stimulated (about 200 msec). The former led to a sensation in the right hand, but this was reported as occurring *after* the sensation in the left hand.

What could be the explanation for this curious effect? Libet had previously argued, on the basis of extensive tests, that when the intensity of the stimulus is very near the minimum threshold required for conscious awareness, it must last for at least 500 msec to be noticed. This is the minimum time required for what Libet calls 'neuronal adequacy'. Thus, he argued, even when threshold stimulation lasts for longer than 500 msec, it cannot actually be registered before 500 msec have elapsed. Libet hypothesized, however, that the brain compensates for this delay by subjectively antedating the experience to a point approximately 500 msec before it was actually registered. The tingling in the left hand was therefore felt to occur before the cortically induced tingling in the right hand. Why was the cortically induced sensation not also subjectively antedated? Because the stimulation of the skin gives rise to a pattern of neural response—the 'primary evoked potential'—that is not found in the case of direct cortical stimulation, and this primary evoked potential appears to be essential for the antedating mechanism to be activated. The whole process, then, looks like this: at 0 msec, the cortex is stimulated, but is not registered as a tingling in the right hand until 500 msec later. At 200 msec, the left hand is stimulated; the response does not gain neuronal adequacy until 700 msec, but the sensation is subjectively referred to 200 msec—*before* the tingling in the right hand. (See Libet 2004 for full discussion.)

This, and related cases, seem to require a rather different mechanism underlying order perception than the one proposed by Mellor. However, this may not be a serious objection to the causal model. In this case, the events occur very close to each other in time, and different mechanisms may then come into play. Where there is a significant gap between perceptions, the suggestion that order perception could be time-insensitive becomes less plausible. The causal model might still be appropriate for standard cases. Interpreting the results of Libet's experiment is in any case not straightforward. It is, for instance, a contentious matter when precisely a perception occurred vis-à-vis another. Dennett, for instance, has argued that there is no particular time at which an experience becomes conscious. Mellor's causal model makes order perception sensitive to the time order of perceptions, but insensitive to its content. One might, however, consider an alternative model: order perception is content-sensitive, but time-insensitive. That is, as Dennett (1991) has suggested, the perceived order may depend on the brain's decision as to what ordering best 'makes sense', given what the perceptions are of.

Rebecca Roache has argued that, once we distinguish between two senses of 'perceive', Mellor's account can be shown to be consistent both with Libet's result and with Dennett's account. To perceive₁ an event is simply for the brain to process information about it, whether or not the mind is consciously aware of the event. To perceive₂ an event is to be consciously aware of it. On Mellor's account, the perceived₂ temporal order of perceived₂ events is determined by the causal order of those perceptions₂. This is entirely consistent with the possibility that the perceived₂ temporal order of events may differ from the temporal order of their perceptions₁. Furthermore, if the perceived₂ temporal order of events arises in certain cases from assumptions concerning their likely order, this cannot be the origin of our concept of temporal order, since such assumptions *presuppose* that concept (Roache 1999: 234–7).

A third, rather different, difficulty is raised by Dainton. He characterizes what he calls the position of the 'memory theorist' in the following terms:

First I hear C. I then hear D, the experience of which is automatically accompanied by a short-term memory-image corresponding to my hearing C. I then hear E, and as I do I have a short-term memory of C-being-followed-by-D. (Dainton 2001: 94)

He then raises the following objection:

. . . the memory theorist is positing a short-term memory of an experience of succession: 'C-being-followed-by-D'. The sort of experience that the memory-account is meant to explain away is in fact being presupposed: we cannot remember what we have not already experienced. If the memory theorist is prepared to admit that we are directly aware of succession when we remember and imagine, why not admit that we are directly aware of succession in ordinary experience? (Dainton 2001: 94–5)

It would be possible to take issue with the principle that we cannot recall what we did not experience. The comments of Baddeley quoted in Chapter 4 point to a picture of memory as something that is constructed from previous information. Given this picture, is it not possible that the construction contains something that was not a feature of any particular experience? However, given the principles of episodic memory endorsed in that chapter, I would pursue a different line. Dainton's memory theorist is someone who thinks that we go straight from experiences of C and D to a memory of C's being followed by D, without an intervening experience of C's being followed by D. But why should the causal influence of the memory of C on the perception of D not give rise to an *experience* of C's being followed by D? Perhaps we do not perceive C's being followed by D in the same way that we perceive C, but we can still have an experience which corresponds in the appropriate way to the later memory.

When it comes to perception of duration, whether of a single stimulus or of the interval between stimuli, the mechanism must be more complex. The considerations of §6.2 lead us to suppose that duration is not the kind of thing that can impinge on us directly. So how do we become aware of it? The answer appears to be that we 'perceive' the duration of an event by *mimicking* it. There is now a considerable amount of evidence supporting the suggestion that organisms sensitive to time have internal time-keepers, or 'biological clocks' (see e.g. Hamer 1968 and Cloudsley-Thompson 1968). Given the huge variety of time-sensitive behaviour in animals, including hibernation, circadian sleep cycles, and locomotive control, it seems likely that there is more than one biological clock governing behaviour. Concentrating on the kinds of case that require fine temporal discriminations, the basic mechanism proposed by *scalar timing theory* involves a neural *pacemaker*, emitting regular pulses, and an *accumulator*, which records the number of pulses emitted by the pacemaker for a given period. Perception of the duration of a given stimulus, according to this theory, involves the following mechanism. Onset of the stimulus causes a switch connecting the pacemaker and the accumulator to close. The accumulator then records pulses until the cessation of the stimulus causes the switch to open, breaking the connection (see Figure 6.1). The accumulator's record may then lead to a judgement of duration, or can be stored in the memory for comparison with other stimuli (Gibbon, Church, and Meck 1984). In one variant of the model, the pacemaker produces pulses at intervals that are variable but whose average duration is nevertheless constant.

Appeal to the slowing down or speeding up of the rate at which the pacemaker emits pulses provides a simple explanation of the familiar fact that time seems to go faster in some contexts and more slowly in others. In some cases, the pacemaker appears to be very seriously altered, with potentially disastrous results. One of the most dramatic instances of this is the *Zeitrafferphänomen* or accelerated time phenomenon. In one case, a patient who had suffered damage to the left prefrontal cortex was driving his car when he suddenly found that objects outside appeared to be rushing towards him at an accelerated rate. Watching television was virtually impossible, because events on the screen were happening too fast to make sense of (Binkofski and Block 1996). One explanation of his experiences is that the damage to the left hemisphere had caused the pacemaker to slow down its production of pulses. In non-pathological

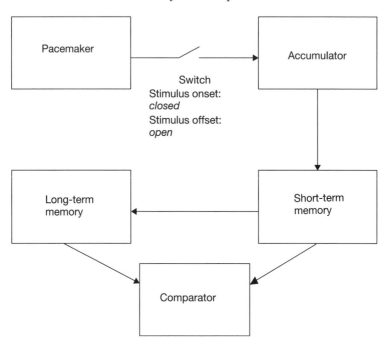

Fig. 6.1: The scalar timing theory of duration perception.

cases, the pacemaker pulse rate appears to be affected, or 'entrained', by external stimuli such as a series of high-frequency clicks (Treisman 1999). The pacemaker may also be affected by drugs, since these produce marked distortions of time estimates (Friedman 1990).

An alternative model for timing involves, not a single pacemaker, but a series of oscillators, whose cycles differ from each other but nevertheless have well-defined phases. Instead of an accumulator recording number of pulses, there is some mechanism that detects, at any one time, what phase each oscillator is in. Thus time from the onset of a stimulus may be represented qualitatively, rather than quantitatively, by the permutations of phases (Church and Broadbent 1990).

The precise details of the mechanism do not matter in the context of addressing our epistemological puzzle. What does matter is the key idea underlying all these models: that of a regular (or averagely regular) neurophysiological process that is effectively a clock system. Thus the perception of time is not simply a passive reception of external stimuli, but an active structuring of stimuli based on an internal system of measurement:

> The ear and eye respond to energies from the environment that impinge upon their receptive surfaces. The tongue and nose sample molecules. To orient us in space, the vestibular system depends on properties the body possesses as a physical object (its response to gravity and inertia). It would seem that the time sense relies on properties of the brain itself as a physical system, the propensity of neurons or neural networks to produce oscillations that can be used as timing devices. Thus one function of the brain may be to act as the sense organ for time. (Treisman 1999: 244)

Controversial though they may be, the models of time perception briefly presented in this section at least give us a starting point in our approach to the epistemological puzzle. What is interesting about these models is that they confirm our suspicion that order and duration do not play a causal role, for on neither model were these temporal properties appealed to as causes of our perceptual states. The challenge, then, is to show how, if time perception is based on something like the mechanisms described by these models, it can lead to knowledge of the temporal properties of events. Since there is a tension between the CTMP and the acausality of order and duration, any resolution of it must pursue one of two strategies: revise our view of what makes perceptual beliefs about intervals true, or revise the causal account of perceptual knowledge.

6.4 FIRST RESPONSE: ORDER AND DURATION AS MIND-DEPENDENT

If purely external relations of precedence and duration are not part of the causal story behind our acquisition of beliefs about them, we might look elsewhere for the truth-makers of those beliefs: in the mind.

6.4.1 An Augustinian inference?

The problem of accounting for time perception greatly exercised Augustine. Here was the difficulty:

> But to what period do we relate time when we measure it as it is passing? To the future, from which it comes? No: because we cannot measure what does not exist. To the present, through which it is passing? No: because we cannot measure what has no duration. To the past, then, towards which it is going?

No again: because we cannot measure what no longer exists. (*Confessions*, Book XI, §27; Pine-Coffin 1961: 270)

What, then, are we measuring? Augustine's answer is that it is a *mental* item: 'It is in my own mind, then, that I measure time.' This is an answer, we might note, that has clear parallels with the sense-datum theory of perception. What some translators have read into his words is the further suggestion that time itself is purely mental. R. S. Pine-Coffin's translation, from which the above passage is taken, follows with the surprising announcement, 'I must not allow my mind to insist that time is something objective'. Other translators have been more cautious, though Henry Chadwick adds a footnote to the phrase 'So it is in you, my mind, that I measure periods of time', comparing Augustine's view with Plotinus's assertion that time belongs to the soul (Chadwick 1991: 242).

Whether Augustine really did intend the inference, our epistemological puzzle provides us with a motivation, not unrelated to Augustine's concerns, for the conclusion that time is mind-dependent. We can offer a parallel argument to Augustine's using the components of our puzzle. So far, we have argued that the putatively mind-independent facts concerning time order and duration are not the causes, direct or indirect, of our corresponding perceptual beliefs. But the CTMP requires that the truth-makers of those perceptual beliefs that constitute knowledge be part of the causal chain leading to those beliefs. Now, although the items whose temporal relations we seem to perceive may be external items (two chimes of the clock, a flash of lightning and a clap of thunder, the beginning and end of a musical phrase), the causes that are responsible specifically for the perception of order and duration are within the mind. So, the neo-Augustinian argument goes, we should look for the truth-makers of the relevant beliefs amongst the *psychological* facts. This would imply, of course, that time order and duration are mind-dependent.

The models of time perception that we looked at in the previous section pave the way for this approach. The scalar timing theory, for instance, suggests a representative theory of perception, where the immediate object of perception is some mental representation of an external state. The mind-dependence strategy simply disposes of the external state (the duration of the event), leaving us with just the representation.

Let us now look in more detail at what this strategy involves—and whether it works.

6.4.2 Prospects for a psychological analysis of time order

Our judgements concerning the temporal order of events depend in part on our spatial relations to those events. On the castle battlements, a cannon is fired to mark the hour of noon. Simultaneously, half a mile away, the town church begins to strike. People walking around the castle walls will hear first the noonday gun, and then the church bell. Those about to enter the church will hear first the bell's initial strike, and then the gun's report. Judgements based solely on observation and in ignorance of one's location vis-à-vis the observed events are therefore prone to error. We can capture this in terms of what I shall call the *Objectivity Constraint*: judgements concerning time order are objectively true or false; they are not simply a matter of opinion, and disagreements between observers are therefore genuine. Any theory of time order must either conform to the Objectivity Constraint, or give us a compelling reason to abandon it. Can the suggestion that time order is mind-dependent be reconciled with the Objectivity Constraint?

At first sight, it might seem that objectivity automatically rules out mind-dependence, but this in general is not so. Consider the case of spatially indexical beliefs, such as 'The centre of the magnetic field is *here*'. This can be given perfectly objective truth-conditions, as follows: any utterance or thought, *u*, of that type, is true if and only if *u* is located at the centre of the magnetic field. Whether or not those truth-conditions obtain is not a matter of subjective opinion. But the 'hereness' picked out by such a token is still mind-dependent, in that it is not an intrinsic property of the external space, but rather a relation between that space and a *representation* of it. There would be no sense to the suggestion that, even in the absence of minds, some place would still be uniquely and absolutely *here*. The truth-conditions for 'here' judgements can both be mind-dependent *and* reflect the perspectival nature of such judgements. Any application of the Objectivity Constraint, therefore, does not automatically and question-beggingly imply mind-independence. Objectivity and mind-dependence are both exhibited by perspectival judgements.

If we represent judgements of time order as similarly perspectival, then we effectively identify actual order with perceived order. To retain objectivity, we have to build in the actual spatial relations between observers and events. Thus, the castle visitors' judgement that the gun went off before the church bell struck was objectively true, given their location, and the church visitors' judgement that the bell struck before the gun went off was objectively true, given *their* location. A consequence, however, of this perspectival treatment of order is that, as it stands, it implies that there cannot be relations of precedence among unperceived events. If this seems too strong, then we could consider the following compromise: we should distinguish between the relation of precedence, on which the direction of time depends, and the relation of temporal *betweenness*, which by itself does not give time its direction. We can treat the former as mind-dependent without treating the latter similarly. So, if we assume that temporal betweenness is independent of psychological facts, all that is required to give direction to the whole time series is the relevant psychological relation between *some* facts in the series. Take the following, as yet undirected, series of events, ordered by the relation of temporal betweenness:

$$e_1 — e_2 — e_3 — e_4 — e_5 — e_6 — e_7$$

Now suppose only e_4 and e_6 in this series are perceived by some observer O, and e_6 is perceived by O as occurring after e_4. Then, given that e_5 is between e_4 and e_6, and e_4 is between e_3 and e_5, it follows that, relative to O, e_3 is earlier than e_5, although neither is perceived. Since time order is supposed on this model to be perspectival, however, the possibility remains that e_6 will be perceived by some other observer as preceding e_4, and this would mean that the entire series would have a different direction for this observer. We do not need to concern ourselves with this consequence, though, since we cannot make precedence perpectival without making betweenness similarly perspectival, so this compromise is not an option. To see this, imagine two observers so situated that for one of them three spatially separated events occur in the order ABC, while for the other they occur in the order BAC. Unless betweenness is perspectival, one of these observers is wrong, thus upsetting the perspectival account of precedence.

Whether or not such a perspectival account is viable, however, it does not help us solve the epistemological puzzle. For it locates the truth-makers of time order judgements not wholly in the mind, but in the relations between external events and observers. And if we try to represent these relations, or facts concerning them, as the causes of our perceptual beliefs, then we face the problems raised in §6.2, and little would have been gained by pursuing the mind-dependence strategy. If, on the other hand, we treat the truth-makers of order judgements as wholly internal to the mind, then we may have restored their causal status, but at the expense of giving up the Objectivity Constraint. Is there a compelling argument for doing this? Well, that it disposes of the epistemological puzzle is an argument, but its force depends on the CTMP having a stronger hold on us than the Objectivity Constraint, and on the absence of any other plausible solution to the puzzle.

6.4.3 Conventionalism about metric

Intuition may favour the Objectivity Constraint in the case of time order, but it is perhaps somewhat less strongly in favour of objectivity when it comes to duration. *Are* there objective facts of the matter as to the metric of time? Consider, for example, two successive swings of a pendulum. Did the second swing take exactly as long as the first? *Objectivism* about temporal metric says that there is a fact of the matter as to whether it did or did not, independently of any means we have of establishing the fact. *Conventionalism* about metric denies this. The truth or falsity of what we might call judgements of isochrony (that two intervals are of the same duration) depends on which clock we adopt as our standard. As Reichenbach articulates it, 'The equality of successive time intervals is not a matter of *knowledge* but a matter of *definition* . . . All definitions are equally admissible' (Reichenbach 1958: 116).

Since the assumption that there are objective facts about metric is a central aspect of our puzzle, conventionalism appears to offer a way out of the difficulty. The human brain (indeed, in many cases, the animal brain) is, according to well-confirmed psychological models, a clock system, and so provides one possible definition of isochrony. If so, then the conventional, clock-relative truth-conditions of perceptual judgements of isochrony will obtain simply by virtue of the psychological processes leading to those judgements. Cause and truth-maker will coincide.

Or will they? They certainly will if Reichenbach is right, for he says that *all* definitions of isochrony are equally admissible. This appears to follow from the conventionalist assertion that no system by which we measure intervals of time is objectively more accurate than any other, since accuracy is defined in terms of one's choice of system. But Reichenbach's position seems unduly liberal. William Newton-Smith suggests that, even for the conventionalist, some judgements about isochrony are not merely eccentric, but false. There are some constraints on selecting a standard: 'if I adopt some deviant clock which gives the ice age, the time between my last two heartbeats and a performance of Wagner's Ring the same duration, I am just wrong' (Newton-Smith

1980: 163). Our standard should be what we could call a *reasonable clock system* (RCS), but how, without invoking objective metric, do we determine what counts as an RCS? One criterion of an RCS would be that it be reproducible, such that different instances of the system tend to remain approximately congruent with each other, and occasional failures of congruence would be readily explicable by means of a simple theory, allowing one to correct errant instances. In addition, the clock system should be compatible with the construction of a coherent physical theory (e.g. of motion). So, having thus defined what it is to be an RCS, the conventionalist can say that judgements of isochrony are true relative to some RCS. The crucial question is then whether or not the human brain constitutes an RCS. If not, then conventionalism can offer an escape route from our puzzle only if we adopt Reichenbach's liberal approach. As a time-measurement system, our biological clock is notoriously variable, and is affected by such things as temperature, excitement and boredom, and of course drugs. Nevertheless, it is a more satisfactory clock system than some systems we might choose as the standard: were it not so, we would be unable to co-ordinate bodily movements as well as we do, or detect subtle variations in the motion of external objects, or be able to appreciate and produce music. Let us not be too particular, then, and let us designate our biological clock an RCS.

Consider now the perception of two successive stimuli by an observer who judges on the basis of their perception that the first stimulus was shorter in duration than the second. The truth of this judgement, according to the conventionalist, is relative to the choice of RCS. So what RCS is relevant here? The obvious answer is: the subject's own internal clock. This, as we saw above, guarantees the causal connection between belief and truth-maker. However, all judgements of relative duration will come out true on this answer, since it is trivial that the deliverances of any RCS are true relative to that same RCS. So we could no longer make the distinction we want to make between those judgements that are accurate and those that are not. If, instead, we choose some other RCS as the standard, then we can distinguish between accurate and inaccurate judgements, since not all deliverances of the biological clock will coincide with those of this other RCS. But the price of this manœuvre is that, except in cases of a special kind, there is no causal connection between the judgement of relative duration and the RCS against which we are assessing the accuracy of the judgement. The exceptions here are provided by certain experimental conditions, such as the following intriguing case concerning instrumental conditioning in rats:

In the standard free-operant procedure, shocks are delivered, in the absence of responding, at fixed intervals . . . and each response postpones the next shock for a fixed period of time (the response-shock or R-S interval). There is thus considerable regularity to the distribution of shocks in time, and it is not surprising that this regularity should at least sometimes be reflected in the subject's behaviour. Although shock can be completely avoided by a rapid and sustained pattern of responding, a slower rate of responding may, if the timing is right, be equally effective. As training continues, some animals learn to avoid an ever-increasing proportion of shocks while emitting fewer and fewer responses, by timing the interval between successive responses to something shorter than the R-S interval. . . .

How is such behaviour to be explained? The obvious point to note is that the probability of responding over time maps the probability of shock. As exposure to the temporal regularities increases, so the subject comes to respond only at those times when the expected probability of shock is high. In the absence of any explicit stimuli, the passage of time since the last response can serve as a signal for the occurrence of the next shock, and also therefore as a signal that another response will cause the omission of this otherwise expected shock. (Mackintosh 1983: 169)

Here an RCS is part of the experimental set-up, and is used to govern the time between certain stimuli. The RCS in question will then both cause and define the accuracy of the subjects' judgements. Thus truth-maker and cause will coincide. In other cases, the choice of RCS will be arbitrary. But is this a problem? After all, for the conventionalist, there is no objective fact of the matter as to whether one interval is shorter than another interval or not, so the subject is not tracking anything. A biological clock is simply a regulatory system, not a means of access to the objective properties of things.

Yet, if we ask for an explanation of our duration judgements, none is forthcoming on the conventionalist view of metric. Or rather, what explanation there is, in terms of psychological mechanisms, seems incomplete. Putting this in the context of the scalar timing model, we might ask: Given that we form a belief about the relative duration of two stimuli on the basis of the numbers of pulses stored in the accumulator, what is it that explains the fact that just *this* number of pulses was collected by the accumulator? The conventionalist has no answer to give. We can push the demand for explanation further and ask why, if duration judgements are only trivially true or contentless, they are so useful. And even in the experimental case described above, where one might explain the rats' behaviour by saying that they are tracking the outputs of some timing device, their success in doing so is mysterious unless one adds that the device in question marks out objectively isochronic intervals.

To sum up the discussion of this section, it initially appeared that treating time order and duration as mind-dependent offered a way out of our epistemological puzzle. If we could locate the truth-makers of our beliefs in the psychological processes leading up to those beliefs, then cause and truth-maker would coincide. This strategy, if successful, would support Augustine's (alleged) contention that the mystery of our ability to perceive time is best answered by taking time to be in the mind. But this strategy appears to lead to an unacceptable subjectivism concerning our judgements—unacceptable not only because it is part of our intuitive conception of time that our judgements have objective truth-conditions (at least in the case of order), but also because the instrumental value of those judgements would be entirely mysterious.

6.5 SECOND RESPONSE: MODIFYING THE CAUSAL TRUTH-MAKER PRINCIPLE

Given the difficulties faced by mind-dependence accounts, we must now ask whether we are *obliged* to take order and metric as mind-dependent in order to solve our puzzle. Is there still room for a view that takes these aspects as mind-*in*dependent? In this section, we shall look at a second strategy for resolving the puzzle, one that involves modifying the CTMP.

6.5.1 Chronometric explanation

Instead of confining the explanatory relationship between perceptual beliefs and their truth-makers to one that is purely causal, we might expand it to include other, non-causal, components of explanation. The modified CTMP becomes:

The Explanatory Truth-Maker Principle (ETMP): Perceptual beliefs that qualify for the title of 'knowledge' have truth-makers that figure in a full explanation of the acquisition of those beliefs.

A full explanation of perceptual beliefs will of course include a causal component. But what other kind of explanation might be relevant? In the case of perceptual beliefs about time order and duration, I propose that the relevant kind of explanation is one that is often assimilated to causal explanation, but should be distinguished from it, a kind I shall dub *chronometric* explanation. Chronometric explanation appeals to the temporal location and extent of things, or to the rate of change. Often it will occur in the context of a causal explanation. Thus, a certain effect may be explained, not simply by the existence of an antecedent cause, but by the location of that cause in time, or by the interval between that cause and another item, or by the rate at which some antecedent change proceeded. Here are some examples of causal explanations that include chronometric explanations:

(a) *Why did the firework explode at* t *?*
Because it was lit five seconds before *t*.

(b) *Why did electricity flow around the system?*
Because the two buttons were pressed simultaneously, thus closing the circuit.

(c) *Why are the two traces on the Campbell–Stokes recorder the same length?*
Because the two intervals of sunshine that caused the traces were equal in duration.

Why should we need to distinguish between purely causal and chronometric explanation? Because, once again, of the difficulties raised in §6.2. The kinds of fact or property appealed to in chronometric explanation do not appear to be causal. This has an obvious affinity with Graham Nerlich's suggestion (1994: 41) that the explanation of the behaviour of moving objects in terms of the geometrical properties of space is a distinctive, non-causal form of explanation: namely, 'geometrical explanation'.

Explanations of, and appeals to, the *rate* at which processes take place are also chronometric. This may seem rather surprising. It implies that we do not, strictly speaking, cause changes in the rate of processes, and also that such changes in rate cannot themselves be the causes of things. We need to be careful how we articulate this point, however. The suggestion is not that changes in rate cannot feature in causal explanations, either as explanans or explanandum, but rather that such explanations must involve a non-causal element. Here are two problematic cases:

(d) *Why did the reaction speed up?*
Because a catalyst was added.

(e) *Why did the ball slow down?*
Because a force was exerted on it.

In both these cases, the explanandum involves facts about intervals. Yet the explanans appears to be purely causal, or at least it seems to involve an event or state that can be given a specific location. What I suggest is happening in these cases is this. It is a brute fact that a given process proceeds at a certain rate. Different processes may proceed at different rates. When some factor appears to be affecting the rate of a process, what it is in fact doing is determining which of a range of related processes is realized. A reaction in the presence of a catalyst is not the same reaction as one in the absence of a catalyst. (A catalyst is sometimes described informally as something which may affect the rate of a reaction without itself being directly involved; but theories of catalysis, whether chemical or physical, all ascribe a much more active role to the catalyst.) So what we have here are two factors: the purely causal fact that the catalyst causes a reaction of a certain type to take place, and the purely chronometric fact that that reaction takes place at a certain rate. Similar remarks apply to the case of motion. Chronometric and geometrical elements both play a role in 'the delay between transmission and reception of the signal was due to the distance involved'.

Having introduced the notion of chronometric explanation, we can now look at the role it plays in the explanation of how we acquire beliefs about order and duration.

6.5.2 How order and duration can be explanatory

Suppose you are waiting at the traffic lights, which are currently red. You now see the amber light appear, followed shortly by the green light. Consider the causal account of your perceptual awareness of the order: you see the amber light, the content of which experience is then stored in short-term memory, which then affects your perception of the green light. The causal order of your perceptions determines your belief concerning the time order of the events perceived. Within this psychological account are two explanations. One is purely causal: the amber light (a) causes your perception of it (P_a); the green light (g) causes your perception of it (P_g), and that second perception is causally affected by P_a. These causal relations, however, are not enough by themselves to guarantee the truth of your belief that a precedes g. So, how does the truth-maker of your belief—that a is actually earlier than g—play any explanatory role in your acquisition of that belief?

First, we need to appeal to the following chronometric explanation: because of the high speed of light and the short distance between you

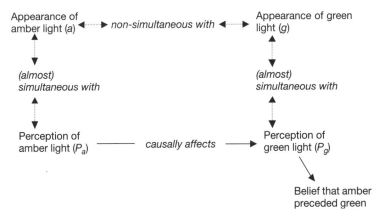

Fig. 6.2: Causal and chronometric elements in the acquisition of belief about time order.

and the events perceived, P_a and P_g are almost simultaneous with a and g respectively. In contrast, a and g are not even approximately simultaneous. These relations are set out in Figure 6.2.

So what is the explanatory connection between truth-maker and belief? Here is one very plausible account: the objective order of the external events, a and g, is what, in part, explains how P_a can causally affect P_g. Because the perceptions are virtually simultaneous with the events of which they are the perceptions, the order of perceptions will mirror the order of events, which is what enables P_a causally to affect P_g, rather than the other way around, which in turn explains why you perceive g as following a. Thus, the truth-maker enters, in a rather simple and obvious way, into the explanation of how you come to acquire your belief.

What of duration? Employing chronometric explanation (at least of the kind that involves reference to duration or rate, as opposed to just simultaneity) clearly implies a commitment to objectivism about metric. That is, chronometric explanation presupposes that there is a fact of the matter as to whether, for instance, one interval is longer or shorter than, or equivalent to, some other interval, even in cases where one interval is not contained within the other. Assuming the legitimacy of this kind of (I submit, non-causal) explanation, we would expect it to be relevant in explaining our perceptual beliefs about metric.

So, let us say that I am presented with two successive auditory stimuli and asked to judge whether or not they were of equal duration.

Suppose that I correctly judge that they are isochronous. According to the scalar timing theory, the onset of the first stimulus closes the switch between pacemaker and accumulator, which then encodes the number of pulses emitted by the pacemaker. Stimulus offset causes the switch to open. The encoded pulses now provide a measurement of the length of the first stimulus, which is stored in the short-term memory. The same process is activated by the second stimulus. The resulting two measurements are then compared, and a judgement of isochrony arrived at. Again, the purely causal facts are not sufficient for the truth of my belief. What more is needed is the chronometric explanation of why just these pulses were recorded by the accumulator. The chronometric explanation will appeal to the objective lengths of the intervals between stimulus onset and offset. These lengths then entail the truth about the isochrony of the stimuli. So, once again, the facts that provide a full explanation of how I come to acquire my belief that the two stimuli are isochronous include the truth-maker of that belief, and the Explanatory Truth-Maker Principle is satisfied.

6.6 CONCLUSION

Let me summarize the results of this long and complex discussion. Psychologists have long recognized that time perception is not directly analogous to perception of objects and their properties, for the temporal features of events do not present themselves for inspection in the way in which spatial properties do. Yet perception does inform us of such things as the time order and duration of events. How does it do this? This psychological problem has a philosophical counterpart: if perceptual knowledge is definable in terms of perceptual beliefs whose truth-maker and cause coincide, how is it possible to have perceptual knowledge of time? For a number of considerations—the problematic nature of time perception, the difficulty of locating temporal properties in space and time, the fact that models of time perception do not include them as causal factors—point to the conclusion that time order and duration are *acausal* properties of events. This is the epistemological puzzle of time perception, and the central aim of this chapter was to solve it.

That the causal theory of knowledge, *prima facie*, runs into difficulties in accounting for certain objects of knowledge is a familiar problem. The most-discussed examples are numbers, as the arithmetical Platonist conceives of them (see e.g. Benacerraf 1965; Steiner 1973; Hale 1987), and moral truths, as the realist conceives of them (see e.g. Mackie 1977). Problems arise in these cases because of the abstract status of such objects. It would, surely, be surprising if difficulties should also arise in a case of *perceptual* knowledge. But time seems to provide such a case.

One approach to the problem is to see it as arising from the assumption that order and duration are mind-independent properties. If instead we look for the truth-makers of our perceptual beliefs about order and duration within the psychological mechanisms underlying our acquisition of those beliefs, this might restore their causal status. Such a strategy, if successful, would vindicate the 'Augustinian' view that, since it is within my mind that I measure time, time is mind-dependent. However, this leads to an implausible subjectivism concerning our beliefs, and no adequate means of explaining either their usefulness or how we acquired them at all.

A better approach is to modify the Causal Truth-Maker Principle. Instead of making causal relations between truth-maker and belief bear by themselves the burden of conferring the status of knowledge, we should require instead only that the truth-makers play *some* form of explanatory role in the acquisition of perceptual knowledge. Explanation, even of a non-logical kind, is not invariably purely causal. And even in what might ordinarily be called causal explanation, there are non-causal components. Within such explanation, beside reference to the causes themselves, there may also be reference to the properties and relations that structure those causes: their location in space and time, for example. It is a question worth pursuing whether a theory of knowledge based on this more inclusive conception of explanation could more easily accommodate knowledge of abstract objects. In the context of time perception, appeal to what I called chronometric explanation helps us solve the epistemological puzzle while retaining our belief in the mind-independence of order and duration. Thus the fact that psychological models of order and duration perception make such perception indirect does not in itself show that these properties are mind-dependent, any more than the indirect nature of depth perception (involving as it does comparison by the brain of the input from the two eyes) shows depth to be mind-dependent.

Before moving on, there is an issue from an earlier discussion that we should revisit. The argument in Chapter 4 against the A-theory made appeal to the Causal Truth-Maker Principle. The A-theorist, it was suggested, could not offer us an account of how the truth-maker of an episodic memory could occur in the causal history of that memory. Now we have suggested moving from the CTMP to the rather wider ETMP, is this earlier argument under threat?

I do not think so. A-theoretical truth-makers of beliefs are subject to constant change. What now makes true my current episodic memory is not what made the original experiential belief true, so cannot be part of any explanation, causal or otherwise, of how I came to acquire that original belief.

Now we have reached the end of Part II, we can answer an enigma posed by Aristotle: 'One might find it a difficult question, whether if there were no soul there would be time or not. For if it is impossible that there should be something to do the counting, it is also impossible that anything should be countable' (*Physics* IV. 14, 223^{a}21; Hussey 1983: 52). *Is* time all in the mind? It depends which aspect of time we are talking about. If order or duration, then no. If A-series position, yes.

PART III

ART AND FICTION

7
Image and Instant: The Pictorial Representation of Time

'How sad it is!', murmured Dorian Gray, with his eyes still fixed upon his own portrait. 'How sad it is! I shall grow old, and horrible, and dreadful. But this picture will remain always young. It will never be older than this particular day of June. . . . If it were only the other way! If it were I who was to be always young, and the picture was to grow old! For that—for that—I would give everything! Yes, there is nothing in the whole world I would not give! I would give my soul for that!'

Oscar Wilde, *The Picture of Dorian Gray*

7.1 GOMBRICH AGAINST THE *PUNCTUM TEMPORIS*

Photographs, paintings, rigid sculptures: all these provide examples of static images. It is true that they change—photographs fade, paintings darken, and sculptures crumble—but what change they undergo (unless very damaging) is irrelevant to their representational content. A *static* image is one that represents by virtue of properties which remain largely unchanged throughout its existence. Because of this defining feature, according to a long tradition in aesthetics, a static image can represent only an instantaneous moment, or, to be more exact, the state of affairs obtaining at that moment. It cannot represent movement and the passage of time. This aesthetic view mirrors a metaphysical one: that change is to be conceived of as a series of instantaneous states. Both of these views were emphatically rejected by Ernst Gombrich in his essay 'Moment and Movement in Art' (1964), which persuasively argues that the very idea of an instant of time is simply a theoretical construct, and one, moreover, that involves both metaphysical and psychological absurdities and prevents us from really understanding what it is that static images represent.

Implicit in Gombrich's argument is a link between depiction and perception. But what is this link, and what role does it play in the argument? *Is* the idea of an instant simply a philosopher's fiction? And if we allow such an idea into our conception of the world, are we thereby committed to a mistaken view of pictorial representation? These are the questions with which we will be concerned in this chapter. We will also attempt to make sense of, and give an answer to, the question of what time span is represented by the static image. I shall begin with a résumé of Gombrich's discussion.

The traditional view of static images is well expressed in two eighteenth-century texts cited by Gombrich. The first is James Harris's *Discourse on Music, Painting and Poetry*, in which Harris writes that each picture is 'of necessity a *punctum temporis* or instant' (quoted in Gombrich 1964: 294). The second is G. E. Lessing's *Laocoon*, which distinguishes between the arts of time and the arts of space. Painting is an art of space because it 'can only represent a single moment of an action' (p. 294) Lessing's distinction, comments Gombrich, 'remained unquestioned in aesthetics' (p. 295), and he suggests that this is the reason for the relative neglect of 'the problem of time' in pictorial representation.

Art appears to offer a way of doing what we cannot do in perception: to freeze a moment of time. But how can we be sure that we really are capturing one of these moments on canvas, rather than presenting a pure invention? Photography, at least when it had developed to the point where only the briefest exposure was necessary, appeared to provide the answer to this question. The famous example of this is Eadweard Muybridge's demonstration (1887), through taking a number of successive photographs of galloping horses, that the position of the legs during the gallop had been systematically misrepresented by painters. This posed a dilemma: should realistic painting aim to imitate photography's capture of a moment in motion, or is art best served by avoiding photographic realism? Gombrich's comments on this take us to the heart of his argument:

Do we not beg the most important question when we ask what 'really happens' at any point of time? We therewith assume that what Harris called a *punctum temporis* really exists, or, more radically, that what we really perceive is the infinite sequence of such static points in time. Once this is conceded the rest follows, at least with the demand for mimesis. Static signs, the argument runs, can only represent static moments, never movements which happen in time. Philosophers are familiar with this problem under the name of Zeno's paradox, the demonstration that Achilles could never catch up with a tortoise and no arrow could ever move. As soon as we assume that there is a fraction of time in which there is no movement, movement as such becomes inexplicable.

Logically the idea that there is a 'moment' which has no movement and can be seized and fixed in this static form by the artist, or for that matter, by the camera, certainly leads to Zeno's paradox. Even an instantaneous photograph records the traces of movement, a sequence of events, however brief. But the idea of the *punctum temporis* is not only an absurdity logically, it is a worse absurdity psychologically. For we are not cameras but rather slow registering instruments which cannot take in much at a time. Twenty-four successive stills in a second are sufficient to give us the illusion of movement in the cinema. We can see them only in motion, not as stills. Somewhere along this order of magnitude, a fifteenth or a tenth of a second, lies what we experience as a moment, something we can just seize in its flight. (Gombrich 1964: 297)

Gombrich's concern, then, is not primarily with the debate over the relationship between photography and painting, but with what that debate presupposed. There are, in fact, no fewer than four theses which Gombrich wants to reject. The first two belong to metaphysics:

(1) There exist instants, i.e. moments of time which have no duration.

(2) Change consists of a series of instantaneous states of affairs.

The next one to psychology:

(3) What we call 'the perception of change' is actually the perception of a sequence of instantaneous states of affairs.

The fourth one to aesthetics:

(4) A static image represents (or more accurately depicts) only an instantaneous state of affairs, not an event covering a period of time.

What is the connection between these theses? Gombrich seems to think that (2) and (3) are consequences of (1), and perhaps also that (3) follows from (2), although he is a little vague about the precise logical relationship. He thinks that (2) actually undermines the notion of change: how could change be composed of changeless parts? We will look at this assertion in the next section. As for (3), it clearly goes against the data. He gives the example of watching a television programme:

When we watch the programme we are, in fact, watching a tiny spot of light traversing the screen from side to side. . . . At each moment of time, therefore, what we really see (if that expression had any meaning) would only be one luminous dot. . . . Actually if we want to pursue this thought to its logical

conclusion the *punctum temporis* could not even show us a meaningless dot, for
light has a frequency. (Gombrich 1964: 297)

Even breaking down our perception of what is happening on the screen
into the smallest units, what we see is a picture, and this is an event
which takes time. So (3) must be rejected. After the discussion of
temporal experience in Chapter 5, we can happily concede this point
to Gombrich. Whether this threatens (1) or (2) is a point taken up in
§7.3. We might note, before moving on, that Bergson at one point says
something highly suggestive of (3):

We take snapshots, as it were, of the passing reality . . . the *mechanism of our
ordinary knowledge is of a cinematographical kind.* (Bergson 1911: 322–3; italics
original)

The snapshots in question, of course, cannot be truly instantaneous.

Gombrich takes the rejection of either (1) or (3) to commit us to
the rejection of (4). Let us first consider the move from the negation
of (1) to the negation of (4). Instants are a fiction, so nothing could
count as representing them. But we have to be careful how we state
this. Paintings (though arguably not photographs) may represent all
kinds of fictions. Gombrich's point, however, is that instants are not
just contingently non-existent; the very idea of them involves absurdity:
the impossibility of change. Well, can pictures not represent the absurd?
Hogarth's *False Perspective* represents an impossible state of affairs, but
we can interpret it because each component of the picture represents
something quite coherent. It is only when we put the components
together that we realize that the states of affairs they represent could not
obtain simultaneously. Escher's staircases and waterfalls are another case
in point. However, if the instant is an absurdity, we cannot explain the
intelligibility of representations of it in such terms. For different parts
of the picture do not represent different parts of the conception of an
instant: if the picture as a whole represents an instant, then *every* part
of the picture, if it represents anything at all, represents an instant. So
if no part by itself represents an instant (since the idea is incoherent),
then the picture as a whole does not.

Next the move from the negation of (3) to the negation of (4). Any
plausible account of how paintings (and sculptures) represent will make
some use of the notion of resemblance: a picture of a sheep will often
look *something* like a sheep. Now resemblance is a perceptual concept, so
there will be some connection between what pictures represent and the
kinds of thing that we perceive, and Gombrich is certainly exploiting

this connection in his argument against the *punctum temporis*. The unstated assumption is that we cannot represent pictorially what we cannot perceive. Once that is admitted, we cannot deny the possibility of perceiving instantaneous states without also rejecting (4). This argument is the subject of §7.3.

Having outlined the shape of Gombrich's attack on the *punctum temporis*, let us turn to his suggestion that it leads us straight to metaphysical paradox.

7.2 CONCEPTIONS OF CHANGE AND THE INSTANT

Even if the idea is ultimately to be rejected, we must have some characterization of an instant in order to assess whether or not it entails any absurdities. What characterization should we give? Gombrich does not provide a formal definition, but one remark is suggestive:

. . . The instant of which the theoreticians speak, the moment when time stands still, is an illicit extrapolation, despite the specious plausibility which the snapshot has given to this old idea. (Gombrich 1964: 303)

The instant, then, is an extrapolation. But an extrapolation from what? On one account, we derive the notion of an instant from a process of dividing an interval into smaller and smaller parts. Clearly, if time is (as we assume) continuous, this process has no end, just as the series 1, $1/2$, $1/4$ . . . has no last member. So if we define an instant as the smallest part of an interval, where a part is itself defined in terms of dividing that interval, we are talking of something that does not exist. The continuity of time entails that there is no smallest part of an interval. If, then, an instant is thus extrapolated, it is indeed an illicit extrapolation.

A quite different conception of an instant does not regard it as a part of an interval at all, but as an indivisible boundary between two parts of an interval. The present moment, for example, may be thought of as a boundary between past and future. This seems to have been Aristotle's view: 'The now is a link of time . . . for it links together past and future, since it is a beginning of one and an end of another' (*Physics*, Book IV, 222^{a}10). As we saw in Chapter 5, Augustine construed this indivisible moment as durationless. So, too, does Aristotle, which is why, presumably, he says later in the same work that 'Time is not composed of indivisible nows' (*Physics*, Book VI, 239^{b}9). That remark is made in response to one of Zeno's paradoxes, and since Gombrich suggests that the notion of an instant leads straight to them, some remarks about them are in order.

Gombrich refers, fleetingly, to two of the paradoxes, the Achilles and the Arrow. Lumping them together as he does obscures some important differences between them. They illustrate, in fact, the difference between the two conceptions of instants presented above. Consider the Achilles paradox. Achilles never overtakes his slower competitor, who has a head start, because by the time Achilles has run the initial distance between them, the tortoise has moved a little further on. When Achilles has covered this further distance, the tortoise will have moved again, by a smaller amount, and so on *ad infinitum*. We can thus represent the race as a series of smaller and smaller steps, each taking a smaller and smaller interval of time. Now if we assume that, for Achilles to overtake the tortoise, this series of decreasing intervals must have an end, a last instant before the overtaking, then we are in trouble. For if time is continuous, there is no such last moment. Between any given moment and the putative moment of overtaking, there is always a third. Consequently, in order to overtake the tortoise, Achilles has to traverse an infinite number of sub-distances, which is impossible.

In the Arrow paradox, it is the second conception of an instant which is operating. The arrow, says Zeno, is at rest at each moment of its flight, and so is at rest throughout the period of its flight. Now if 'moment' here meant even an infinitesimal interval of continuous time, then there would be no justification for denying that the arrow moved during that interval: even an infinitesimal interval has earlier and later parts, and so can contain motion from *that* place to *this*. So the paradox depends on treating the moment in question as genuinely indivisible, not composed of earlier and later parts. It might be durationless, or, if time is only finitely divisible, then we could conceive of each instant as having a very short (but not infinitesimal) duration. In this second case, there would be no need to assert, with Aristotle, that 'Time is not composed of indivisible nows'. The essential point about each instant, whether or not it has a duration, is that it be indivisible, for then it cannot contain motion.

Zeno's paradoxes, of course, are well known and have generated a considerable literature. (For sources and discussion see Barnes 1982 and Sorabji 1983.) This is not the place for a detailed assessment, especially as the focus of this chapter is pictorial representation, and the paradoxes do not form the main component of Gombrich's attack on the *punctum temporis*. But, in the context of this discussion, it is important to ask whether it is the notion of the instant that leads to the paradoxes. As far as the Achilles is concerned, instants do not need to enter the discussion at all. The central idea that generates the problem is that of intervals that can be divided indefinitely. Indeed, if we were to replace continuous time with discrete time, in which an interval consists of a finite number of moments, the Achilles would disappear (provided we treated space as discrete, too).

The notion of the instant is, however, essential to the Arrow paradox. There must be an indivisible instant, 'the moment', as Gombrich puts it, 'when time stands still'. As we have seen, this need not be, as Gombrich assumes, durationless. But one standard response to the Arrow is to invoke the very conception of change that Gombrich thinks leads to the problem in the first place, namely thesis (2): Change consists of a series of instantaneous states of affairs. Russell calls this the 'static' account of change (Russell 1903: 350). As applied to the case in hand, the static account holds that something is in motion if and only if it occupies different positions at different times. We can further distinguish between moving *in* an instant, which we can concede is impossible, and moving *at* an instant. An object is in motion *at* a given instant if the object is in a different position at any instant before or after that instant. Properly understood, the static account of change, far from leading us to Zeno's Arrow, offers a way out of it.

There is a further move in the dialectic, urged by Lear (1981): intuitively, when something moves, its motion is *present*. If we interpret this merely as moving *at* the present moment, in the above sense, then we make a present state of affairs (the motion) derivative of states of affairs at other moments, and this seems just wrong. For surely what happens *now* is ontologically fundamental, reflecting the ontologically privileged nature of the present. If we are presentists, then the point will be even more significant. If only the present is real, and the present is, as Augustine taught us, indivisible, then on the static account motion cannot be a fundamental aspect of reality. (See Le Poidevin 2003: ch. 9 for discussion.) A-theorists thus have reason to reject the static account. What will they replace it with? Change, they may say, is more than mere displacement along some dimension (position, pitch, temperature, etc.), but is an inherently dynamic state. Presentists, at least, will need to hold that it is an instantaneous dynamic state, for only such a state could be intrinsic to the indivisible present. The idea of an instantaneous dynamic state is not obviously absurd, and has attracted serious support (e.g. Tooley 1988; Bigelow and Pargetter 1989).

Now perhaps Gombrich sees change in these dynamic terms, but if so, his argument is more with (2) than it is with (1). The notion of an instant is, as we have seen, compatible with both conceptions of change.

7.3 DEPICTIVE AND NON-DEPICTIVE REPRESENTATION

We now turn to Gombrich's main argument, that the idea of an instant is 'psychologically absurd'—i.e. that we cannot perceive instantaneous states, and that, in consequence, static images do not represent instantaneous states.

As noted in §7.1, Gombrich seems to suggest that (3), the psychological thesis that our perception of change is really the perception of a series of instantaneous states, follows from (2) the metaphysical thesis that change just *is* a series of such states. Does it? (3) is in fact ambiguous. It could mean: that which in the world corresponds to the change we perceive is a series of instantaneous states. If it does mean this, then it is just another way of stating (2): it is a thesis about the world, rather than about our representation of it. But (3) could also mean: we perceive individual instantaneous states, and a series of such perceptions adds up to the 'perception of change'. This, I think, is how Gombrich interprets it, but so interpreted, the inference from (2) is problematic. His thinking may be this: if change is a series of instantaneous states, then in perceiving change, we must perceive such a series. But if we cannot perceive individual instantaneous states, we cannot perceive any part of this series, and if we cannot perceive any part of the series, we cannot perceive the series itself. Thus, we cannot perceive change. This is absurd, so the conception of change stated in (2) and implied by (1) must be wrong. If this is his thinking, however, it is fallacious. To perceive something is not necessarily to perceive all its parts, even if all those parts make a causal contribution to the perception. An instantaneous state may impinge on us without our registering the fact, but if it and similar states had not impinged on us, we would not have registered the series of states as a whole.

Now to the crucial step, the move from the rejection of (3) to the rejection of (4), that static images represent only an instant of time. To assess this move, we need to be more explicit than Gombrich is (at least in his 1964 paper) on the connection between perception and pictorial representation.

First, we need to pay some attention to a distinction drawn by Gregory Currie and others between depictive and non-depictive representation. Intuitively, depiction is representation by means of visual resemblance. A picture of a sheep depicts a sheep by visually resembling a sheep (in certain respects). But pictures represent more than they depict, by virtue of what they depict:

A cinematic image may represent the man it depicts as sad or angry. These qualities are not depicted, because sadness and anger are not qualities that could be depicted; they are mental qualities not accessible to vision or the other senses. But often this nondepictive representation is strongly tied to what the work does depict. Usually, the representation of the nonperceptual quality of sadness occurs in virtue of the depiction of visible qualities; the man is represented as having a sad expression. (Currie 1995: 91)

Depiction, then, is limited to the kinds of things that are more or less immediately accessible to vision, which in the case of static images seems to limit the features depicted to spatial properties and colours. We may, on the basis of these features, recognize various qualities in what is depicted, but only by a process of (perhaps unconscious) inference: e.g. *those* features of a face are associated with *that* emotion.

Given this distinction between depictive and non-depictive representation, the door is open to the view that static images may represent aspects of time that they are unable to depict. Consider the strip cartoon. A sequence of relevantly similar (but also relevantly different) pictures in a linear sequence may represent the passage of time by virtue of the convention that pictures on the right represent events which are later than those represented by pictures on the left. Another method of representing time and change is illustrated by those futurist paintings in which representations of non-simultaneous states are superimposed on each other: Duchamp's *Nude Descending a Staircase*, Balla's *Dynamism of a Dog on a Leash*, or Malevich's *The Knife-Grinder*. And a familiar technique in photography is the long-exposure photograph, where moving objects appear blurred against a clear and therefore static background. All these are static representations of motion, and hence of intervals of time.

So if Gombrich's target is the thesis that static images represent (by whatever means one cares to mention) only an instantaneous state of affairs, then it is a relatively easy one to hit. But the eighteenth-century authors he quotes seem in any case to be aware of painting's ability to represent, in a wide sense, more than a single moment. Take, for example, this remark from Anthony, Lord Shaftesbury's *Second Characters, or the Language of Forms*: 'It may however be allowable, on some occasions, to make use of certain enigmatical or emblematical devices, to represent a future time' (Shaftesbury 1914/1713: §8). And the representation of past time can be quite unenigmatical:

How is it . . . possible, says one, to express a change of passion in any subject, since this change is made by succession; and that in this case the passion which is understood as present, will require a disposition of body and features wholly different from the passion which is over, and past? To this we answer, That notwithstanding the ascendancy or reign of the principal and immediate passion, the artist has power to leave still in his subject the tracts or footsteps of its predecessor; so as to let us behold not only a rising passion together with a declining one, but, what is more, a strong and determinate passion, with its contrary already discharged and banished. As, for instance, when the plain tracts of tears new fallen, with other fresh tokens of mourning and dejection, remain still in a person newly transported with joy at the sight of a relation or friend, who the moment before had been lamented as one deceased or lost. (Shaftesbury 1914/1713: §10)

Let it then be conceded on all sides that static images can represent, by some means or other, more than the instant. The interesting question is whether such images can *depict* more than the instant. Now if depiction is representation by resemblance, where this requires a sharing of properties between representation and represented, then it seems that they cannot. Plays and films can represent time *by* time: temporal order in what is represented can be reflected in the temporal order of the representation. Such representation is, to use Currie's term, *homomorphic*. The representation of time in static images, by contrast, is *heteromorphic*: other times are represented either spatially, or by calling to mind, by devices such as the one mentioned by Shaftesbury, the past and perhaps also future context of the moment on canvass. Lessing's dichotomy, then, survives in some form: the arts of time are those that can represent time and change homomorphically; the arts of space those that can represent time and change only heteromorphically.

This does not settle the issue, however, for there is reason not to identify depiction with visually homomorphic representation. Paintings can depict whilst sharing very few features with their subjects. A painting or drawing of a cow is two-dimensional, typically not life-sized, and may lack colour. It may present only part of the cow, or in very abstract terms. The spatial relations of the parts of the cow may be greatly distorted or exaggerated. Yet there is something about the picture that powerfully conveys the idea of a cow. A satisfactory theory of depiction must therefore explain the sense in which pictures are *like* their objects, while at the same time sharing so few properties with those objects.

Currie's suggestion, building on the ideas of Flint Schier (1986), is that an image depicts an *x* by virtue of the fact that it triggers, by means of its visual features, an *x*-recognition capacity in the observer. This captures the sense in which looking at a depiction of a cow is somewhat like looking at a cow, even though picture and cow share very few intrinsic features. The relevant shared feature on this account is an *extrinsic* property of the depiction and its subject: its disposition to induce a certain response in us. The triggering of the *x*-recognition capacity is something quite immediate, not requiring conscious judgement, so there is no need to say that the picture of the cow gives us the *illusion* of being presented with a cow. As Currie points out, this account explains a feature of our understanding of pictures stressed by Schier, its 'natural generativity'. Let us allow that all but the most realistic pictures may require *some* grasp of convention. Once I have mastered the ability to recognize a picture as a depiction of a cow, I can recognize similar pictures as depictions of anything else, such as a horse, or a tree, provided only that I am able to recognize a horse, or tree, etc. Language, by contrast, is not naturally generative. Learning to interpret a word in, say, Anglo-Saxon, as the word for 'wine', does not bring with it the ability to interpret other Anglo-Saxon words for familiar objects. (See Schier 1986: 43–4; Currie 1995: 80–90.) The reason for this difference is that pictures, unlike words, exploit perceptual capacities that we already possess.

Of course, objects may quite accidentally trigger certain of our feature-recognition capacities, but we would not necessarily want to call those objects depictions. The appearance of a cloud, for instance, may make us think of a flying saucer, but the cloud is certainly not a depiction of a flying saucer. We would need to build into the account offered here the idea that the triggering is non-accidental, then, perhaps appealing to a Gricean account of meaning. But we need not worry too much about providing sufficient conditions, because our aim here is simply to draw the distinction between depictive and non-depictive representation, not to offer a complete account of what it takes for something to represent.

Some accounts of depiction tend to focus on the depiction of objects, and cannot easily be generalized to account for the depiction of, for example, properties, relations, or states of affairs. (See e.g. the account given by Christopher Peacocke (1987), in which depiction is explained in terms of the spatial properties which the representation has in the visual field.) An advantage of Currie's account is that it can be generalized in this way, for we may appropriately talk in terms of a recognition capacity for properties, relations, states of affairs, and indeed changes in any of these things.

This account brings into the open what plays such an important role in Gombrich's argument: namely, the connection between depiction and perception. The central insight is the idea that we cannot understand depiction just by comparing objects and pictures: they are related to each other by virtue of their relation to a *perceiver*. Perceiving objects and perceiving depictions of them have something in common, and this something is what explains how depictions depict. We can now see why Gombrich's inference from the impossibility of perceiving the instant to the impossibility of depicting the instant can be justified: if we do not possess a recognition capacity for instantaneous states, they cannot be depicted.

Once we have the depictive/non-depictive distinction in place, we can define pictorial realism in terms of it. A picture is realistic to the extent that the features it represents are predominantly represented depictively, rather than by other means. Realism is thus a matter of degree (Currie 1995: 90–1). But Currie does not view depiction as a matter of degree: for him, a feature is represented either depictively or non-depictively. The question then is where to draw the line. We considered above Currie's assertion that, whereas the spatial features of a face associated with sadness could be depicted, the sadness could only be represented non-depictively. The account of depiction as triggering recognition capacities, however, is somewhat in tension with this. For we *do* possess a sadness-recognition capacity, and that, surely, is, triggered by pictures of sad faces.

Currie suggests two quite different mechanisms for feature recognition: one involves reasoning and reflection, general beliefs about what is probable based on previous experience; the other is 'more automatic, less flexible, less rational' (p. 85). The kind of recognition capacities that define depiction are, he holds, of the latter kind. But it is not obvious to me that the recognition of emotions is not quite immediate and instinctive (though not altogether innate): it is certainly something we see in animals. On the other hand, if we are going to be very strict about the level of cognitive sophistication involved in recognizing depictive features, then we should perhaps refrain from saying even that an image can depict a *man*, or a *cow*, as opposed to the characteristic spatial features of a man, or a cow. And what about representation of the third dimension? Can solid, three-dimensional objects be depicted, or only their two-dimensional surfaces? The representation of depth involves some interpretation. Of course, perspective in pictures is meant to mirror perspective in depth perception, but we do not, typically, view pictures stereoscopically, so one of the most important guides to depth available in perception (the difference between the two retinal images) is not normally available when we look at pictures.

We avoid demarcation disputes by allowing that the difference between the lower-level, less rational recognition capacities and the higher-level, more consciously interpretative ones is one of degree rather than kind. What we have is a continuum of capacities, involving varying degrees of cognitive sophistication. We can then arrive at a modification of Currie's account, as follows: to the extent that an image triggers a recognition capacity for x that is at the 'less sophisticated' end of the continuum, we will be more inclined to say that the image depicts an x. To the extent that the relevant capacity is nearer the 'more sophisticated' end of the continuum, we will be more inclined to say that the image non-depictively represents an x. This does represent a significant departure from Currie's account, since it makes the depictive/non-depictive distinction a matter of degree: the extent to which a feature can be said to be depicted varies inversely with the degree of cognitive sophistication required to recognize it.

Now that we have what I hope is a plausible theory of depiction, we can look at the effect it has on Gombrich's argument.

7.4 DEPICTION, CHANGE, AND THE CINEMATIC VIEW

Can static images depict change and movement? Employing the account of depiction arrived at in the previous section, this question becomes: Can static images trigger our change and movement recognition capacities at a level that does not require a relatively high degree of cognitive sophistication?

If we set the threshold for the relevant degree of sophistication very low, and insist that images only depict motion, say, if they trigger our normal perceptual motion-detection capacities, then an image depicts motion only if we see it as moving. But, with a few interesting exceptions, we obviously do not see a static image as moving. The exceptions are images such as Bridget Riley's *Cataract 3*, or *Fall*, which create the illusion of movement. Gombrich ends his paper with the following comment on Riley's pictures:

[E]ven abstract art can elude the static impression at least in those extreme cases which exploit fatigue and after-images to produce a sensation of flicker and make the striations and patterns dance before our helpless eyes. An explanation of these phenomena experienced before in the black and white paintings by Bridget Riley . . . is still being sought, but even the first attempts throw a fascinating light on the complexity of the visual process. Experiments such as these are wholesome reminders of the inadequacy of those *a priori* distinctions in aesthetics which were the subject of this paper. (Gombrich 1964: 306)

On the account of depiction that we have been invoking, can we say that Riley's images depict movement? That what is being triggered in these cases is exactly the same kind of process as goes on when we really do perceive movement is not an uncontroversial matter. According to one hypothesis, such images stimulate the retinal movement-detector cells as a result of tiny and involuntary movements of the eyeball. But this is not the only view. Such cases are all the more complex since, paradoxically, the images appear to move and stay still at the same time. (Cf. the discussion of the waterfall effect in Chapter 5.) In so far as these images are triggering certain motion-detecting mechanisms, they are not triggering others (such as the system responsible for detecting change in location), hence their paradoxical nature. In any case, their abstract nature means that it hard to see them as depicting a changing *object*.

As I say, these images are the exception. Most static images, at least, do not induce the impression of actual movement. So what impression do they induce? Consider again Bergson's 'cinematic' view of perception: we take, as it were, a series of snapshots of the world, and the series as a whole leads to an impression of movement or change. If the snapshots are supposed to be literally instantaneous, then, as Gombrich points out, the thesis is false. But the cinematic view does not depend on this. Bergson says that the snapshots are *almost* instantaneous. The point is that they are psychologically instantaneous: they register only a state, not a duration. Gombrich would probably want to reject this thesis, too, and argue that the experience of the world is essentially dynamic, that the division of experience into snapshots makes no sense. We can reply to this by appeal to the resolution of the phenomenological paradox discussed in Chapter 5. Each perception presents its content as present, and we cannot discern earlier and later elements in the present. However, the causal influence of earlier perceptions on later ones gives rise to the experience of pure succession. Thus, the cinematic view of perception is quite consistent with the perception of succession as present. The cinematic picture also suggests a view of what static images depict: if the perception of motion consists of a series of perceptual 'stills', then static images depict the part of motion that we register in one of these 'stills'.

Now let us raise the threshold of cognitive sophistication a little further, and consider a particularly dynamic painting such Butler's *Charge of the Scots Greys*, representing a cavalry regiment charging into battle on horseback. The elements represented are clearly in motion: the galloping horses, the brandished swords, the soldiers shouting to each other, some nearly falling off their steeds. Nothing could be further from a still-life. But does it *depict* movement? It certainly seems that there are some features of static images that, while not giving rise to the actual sensation of movement, nevertheless strongly suggest it, in ways that do not involve conscious interpretation. Gombrich draws attention to the fact that, when figures are portrayed in an asymmetric or unstable position (i.e. one which they would not be able to sustain for any length of time), the suggestion of movement is more powerful (1964: 303–5). In these cases, we might be prepared to say that the representation of movement is closer to the depictive end of the continuum.

There is a parallel, we might suggest, between the depiction of change and the depiction of emotion. As we saw earlier, Currie wanted to insist that an image can only represent emotion non-depictively, by depicting certain physical features of a facial expression. He might, then, want to say that a painting like Butler's non-depictively represents movement by depicting certain unstable spatial relationships. But, as I suggested, a face can be recognized pretty immediately as sad, and the degree of cognitive sophistication required to recognize a depicted face as sad is not therefore sufficiently high for us to deny that this is genuinely the depiction of emotion. We should draw a similar moral, I suggest, concerning change.

It is time to draw the threads of this discussion together. I am in substantial agreement with Gombrich's conclusion, but not with all the details of his route to that conclusion. In particular, he does, I think, overstate the case against the '*a priori* distinctions' concerning time and pictorial representation: namely, the distinction between an instant and an interval and that between the arts of time and the arts of space. Those distinctions *are* useful in interpreting pictures. The notion of an instant, when properly characterized, does not obviously involve us in logical absurdities; nor does it entail, or even suggest, dubious theses about perception or depiction. And the arts of time/arts of space distinction can be reconfigured as the distinction between those arts that represent time homomorphically and those that do not. There is another useful, though less clear-cut, distinction: that between depictive and non-depictive representation. Having understood it, we can see that the answer to the question of what time span a static image represents depends on the level of cognitive sophistication involved in responding to the features of that image. At the most basic level, static images may depict a part of an event, whatever could be taken in at a single glance. At a much higher level, involving an appreciation of dramatic context, they suggest a much longer time span. Somewhere between these levels, certain spatial features of the image may convey the impression (though not literally the sensation) of change or motion. There is no incompatibility between these answers. An image may represent both an interval and smaller constituents of that interval, precisely because it can represent its objects in different ways.

To conclude: the static conception of change, and the corresponding cinematic view of perception, need not obscure our understanding of the pictorial representation of time.

8

The Fictional Future

'Before I draw nearer to that stone to which you point', said
Scrooge, 'answer me one question. Are these the shadows of things
that Will be, or are they the shadows of things that May be, only?'

Still the Ghost pointed downward to the grave by which it
stood.

Charles Dickens, *A Christmas Carol*

8.1 FATE, FICTION, AND THE FUTURE

Macbeth, Act I, Scene v: Lady Macbeth is reading a letter from her
husband, in which he recounts his dramatic meeting with the three
witches on the heath. They hailed him, not only by his present title of
Thane of Glamis, but also, bewilderingly, as Thane of Cawdor and as
'King hereafter'. No sooner had they left him than Macbeth received
the news that the treacherous Thane of Cawdor had been executed, and
that the King, Duncan, had bestowed the title on him. What could it
all mean? Pondering the mystery, Lady Macbeth knows at once what
her course of action must be:

> Hie thee hither,
> That I may pour my spirits in thine ear,
> And chastise with the valour of my tongue
> All that impedes thee from the golden round,
> Which fate and metaphysical aid doth seem
> To have thee crown'd withal.

Yet her thoughts betray some conflict. On the one hand, she is inclined
to take seriously the witches' prophecy of her husband's meteoric rise
as an authoritative statement of future events—it has, after all, already
been partly vindicated. But, on the other, she fears that it may not come

to pass, for she recognizes her crucial role in persuading Macbeth to take steps he would shrink from if left entirely to his own devices.

Now, it might be said, there is no real conflict here. For, unless we are fatalists, we understand that the determinateness of the future is not incompatible with our affecting it. So, even if it is already the case that Macbeth will be King of Scotland, that state of affairs is still causally dependent on current actions. However, Lady Macbeth seems prey to genuine doubts:

> [Thou] shalt be
> What thou art promised. Yet do I fear thy nature;
> It is too full o' the milk of human kindness
> To catch the nearest way.

And they are fully justified, for at a crucial moment when Duncan is within the walls of their castle Macbeth suddenly announces, 'We shall proceed no further in this business'. For a while it seems that future things, far from being determinate, remain as yet mere possibilities.

It is this conflict between two views of the future, as *fixed* (i.e. determinate, as opposed to causally determined) and as *unfixed* (indeterminate), that is the central concern of this chapter, and I shall explore it in the context of fiction. The thesis I want to examine is that some fictions, at least, represent the future as fixed. I take this to be captured in the following proposition:

The Fixed Fictional Future (FFF) thesis: Some fictions are such that some future-tensed statements are true in those fictions.

If we already grant the notion of truth-in-a-fiction (something I shall not question here), then the FFF thesis is a plausible one, and becomes more so when we look at specific instances. It turns out, however, that the thesis generates a surprising number of difficulties when we try to accommodate it within a theory of truth-in-fiction. Our attempt to do so will, I hope, shed some light on the nature of fictional time, and also on the relationship between fictional truth and the imagination.

What, then, is the motivation for the FFF thesis? At various points during Act I, we can intelligibly ask, 'Will Macbeth be King of Scotland?' We do not have to wait long for our answer. As in real life, so, we suppose, in fiction: what is sometime future becomes present and then past. We have the powerful sense that his becoming King is inevitable, that it was already true at the time of his encounter with the witches, and perhaps even before that. Lady Macbeth's reference to 'fate and metaphysical aid' reinforces that impression. We could perhaps argue that the witches have, not knowledge of the future, but merely justified belief, the justification being that announcing to Macbeth that he would be King was highly likely to bring about the state of affairs it foretold. But even if we could present this as just a piece of deliberate mischief on the witches' part, rather than a genuine vision of the future, their later predictions (that the King will not be defeated until Birnam Wood comes to Dunsinane, and only then by one not born of a woman) could hardly support such a construction. Although their utterances are open to misinterpretation, and are indeed misinterpreted, the witches evidently know what fate lies in store for the hapless murderer of Duncan.

Of course, not every statement we might care to make about the fictional future will have a truth-value. What is given in fiction under-determines many propositions, such as whether Macbeth is left-handed. The FFF thesis says only that, for *some* fictions, *some* statements about the fictional future can plausibly be ascribed a determinate truth-value.

Another example of fictional fate is provided by the myth of Oedipus, whose father Laius is told by the oracle that the child conceived through his union with Jocasta will be his murderer. As with Macbeth, the prophecy is one of the causes of its fulfilment: in attempting to avoid his fate, Laius takes steps which—unhappily—ensure that Oedipus will not know his father until it is too late.

But not all visions of the fictional future are part of the fiction. In J. B. Priestley's *Time and the Conways*, we are presented in Act I with a comfortable and apparently cheerful middle-class family whose younger members seem well-favoured and justified in their hopes for the future. In Act II we see the same group of people, twenty years on, disappointed and disillusioned. Act III returns us to the time in which the play began, and we see the youthful Conways in a different light. There is still the same air of hope, and the banter goes on as before, but this time we see the seeds of their future misfortune in the defects of character which are just beginning to emerge and which so dominate the later picture. We know, as we watch the final act, what will happen, since we have already seen it. It is we, rather than the fictional characters, who are able, as Banquo puts it, to 'look into the seeds of time | And see which will grow and which will not'. Here the fictional future is entirely fixed, at least within the narrow compass of the fortunes of this particular family.

Our final, and rather obvious, example is H. G. Wells's novel *The Time Machine*. The time-traveller journeys to the remote future, and is not merely a passive witness of that future, but an agent of change within it. Indeed, it is this causal interaction that is the source of his knowledge of the future. For this to be possible, the future (i.e. the future in relation to the time of his departure) must be determinate. Moreover, it must be determinate irrespective of whether the time-traveller actually visits the future or not, for we could hardly suppose that it is his arrival in the future that makes it determinate—consider all the times through which he passes without alighting.

But what of cases where neither we nor any fictional characters are vouchsafed a vision of the future? Provided a fiction is temporally structured—that is, is one in which events are represented as exhibiting a temporal order—then, as it unfolds, certain future-tensed questions raised at earlier stages receive determinate answers. Does this guarantee that earlier future-tensed statements have a determinate truth-value? That would be too strong a conclusion to draw at this point. For, like fiction, history unfolds, in that certain states of affairs come to pass; but it does not follow that it was *always true* that those states of affairs would come to pass. The future may be genuinely indeterminate, in a sense to be explained more fully in §8.3. So, by analogy, the mere fact that fictions unfold seems perfectly compatible with the unfixity of the fictional future. On the other hand, most fictions are what we might call 'complete' fictions: that is, ones whose content is entirely fixed before we engage with them. Does this not suggest that the truth-value of any statement we might make at any stage in the fiction is already fixed? Not necessarily. What is true *in* the fiction is one thing; what is true *of* the fiction, another. Suppose I wish to ruin my friend's enjoyment of a film she has never seen before. As we watch the murderer tiptoe upstairs, rope in hand, to the room where the defenceless and bedridden old lady is lying, unaware of the terrible fate which seems to be awaiting her, I say, 'It's all right: the police will arrive in the nick of time.' What I say is true of the film, and the celluloid frames that prove it so are waiting their turn to go into the projector. It does not follow that *within the fiction* it is determinately true at this point that the police will arrive. In any case, not all fictions have their contents fixed before we engage with them: consider *interactive* fictions, where we in part determine fictional content. Surely here, at least, the fictional future is genuinely open.

Later, I shall argue that there may be very little to separate interactive fictions from complete fictions, but at this stage the thesis I want to pursue is a relatively modest one: in so far as there is, for any temporally structured and complete fiction, a fictional past, present, and future, the content of some fictions of this kind suggests—for example, by explicitly invoking the notion of fate—that there are future-tensed statements ('Birnam Wood will come to Dunsinane', 'Oedipus will kill Laius', 'The Conways' hopes will be dashed', 'The Eloi will defeat the Morlocks') which are determinately true in those fictions. This is the FFF thesis, or at least a qualified version of it, and it is less the deliverance of some theory than an intuition against which we may test theories of fiction. But intuitions themselves are not immune from testing, and if they prove to be sufficiently problematic, theoretically, they may be abandoned.

Before we consider the FFF thesis in detail, I want to examine two general issues concerning fiction's capacity to represent the future.

8.2 THE FICTIONAL A-SERIES

Talk of what is *going to happen* in a fiction, we have supposed, is entirely coherent. And why not? After all, can we not talk of what *has* happened, and what is happening *now*? Recall McTaggart's distinction, introduced in Chapter 3, between the A-series (past, present, and future times) and the B-series (earlier and later times). It is, it seems, a presupposition of the FFF thesis that there is a fictional A-series, a fictional past, present, and future, as well as a fictional B-series. But McTaggart himself casts doubt on this supposition. He imagines someone making the following objection to the view that the B-series is dependent upon the A-series:

The . . . objection rests on the possibility of non-existent time-series—such, for example, as the adventures of Don Quixote. This series, it is said, does not form part of the A-series. I cannot at this moment judge it to be either past, present, or future. Indeed, I know that it is none of the three. Yet, it is said, it is certainly a B-series. The adventure of the galley-slaves, for example, is later than the adventure of the windmills. The conclusion drawn is that an A-series is not essential to time. (McTaggart 1927: 26)

His brief response to this is that 'Time only belongs to the existent'. Fictional events do not form an A-series, but they do not form a B-series either; so fiction tells us nothing about the actual relationship between the A-series and the B-series.

This is a bit quick. McTaggart's view, after all, is not just that B-series positions are determined by A-series ones, but also that B-series terms are *definable* in A-series terms (see his analysis of 'earlier': (1927: 271).

Whether or not fictional events are real, they can still be described using B-series terms, as the above example shows. And if B-series terms are definable in A-series terms, then any B-description (i.e. a description involving B-series terms) can be translated into an A-description, such as, for example, 'when the adventure of the windmills is present, the adventure of the galley-slaves is past'. If we cannot make sense of the components of this sentence, as McTaggart seems to concede, then this casts doubt on his definability thesis.

As we saw in Chapter 3, McTaggart is not the only proponent of the definability thesis: Arthur Prior, Michael Dummett, Richard Gale, Peter Geach, John Lucas, and Roger Teichmann have all defended it. So the analysis of fictional time has significance beyond our understanding of fiction.

Why should anyone think that we cannot judge a fictional event to be, for example, present? Perhaps because the implication would be that the event in question is real. But we can surely avoid that implication by enclosing the sentence describing the event within the fictionality operator: 'It is true in *F* that *e* is present.' The problem, surely, is this: we have to locate some temporal point within the fiction from which we can ask what will happen, for questions posed in tensed terms, of what *will* happen in the fiction, or what *has* happened or is *now* happening, can be posed intelligibly only from some temporal perspective. (Whether this is true outside the fictional context is discussed in the next chapter.) If we do not locate such a point, but take a God's-eye view of events, perhaps for the purposes of literary analysis, we can surely only talk in B-series terms of what occurs before what. Now one line of thought has it that, since we are not in fact located within fictional worlds, and so do not have a temporal perspective on them, we cannot coherently talk of what is past, present, or future within a fiction (see e.g. Le Poidevin 1988). But, as Gregory Currie (1998) has pointed out, the premiss of this line of thought is at odds with the role of make-believe in fiction. Some fictions, and perhaps, if Kendall Walton (1990) is right, all fictions, just *are* games of make-believe. If I am engaged in a game of cops and robbers with my children, I can make-believe—so making it fictionally true—that I shall shortly capture one of the robbers and put him in prison. And such fictional self-location need not be so energetic. I can simply be watching a performance of a play, and identify what is now happening on-stage with what is now happening in the fiction, and I may go on to anticipate what is going to happen. Imagine the scene: someone is left all alone in the drawing-room of an isolated country mansion. Suddenly, the lights go out. Surely, at any moment now there will be a scream, and when the lights go back on, there will be a lifeless body on the floor.

It is a common view in aesthetics that what I see on-screen or on-stage is fictionally present. Currie calls this 'the claim of presentness' (1995: ch. 7), and identifies it in the writings of a number of film theorists. The plausibility of the claim of presentness perhaps lies in its connection with another plausible thesis: that when we watch the representation of fictional events, we imagine seeing the fictional events themselves. So when we actually see Ian McKellen greet Judy Dench, we imagine seeing Macbeth greeting Lady Macbeth. And if what we see, we see as present, then when we imagine seeing this fictional scene, we imagine it as present. And so we arrive at the claim of presentness. We can extend this to reading a novel: when I read the description of a fictional event, I imagine observing the event, or perhaps listening to the narration of it. The idea, then, is that the fictional A-series can be preserved by the imaginative projection of oneself into the fiction.

Two difficulties arise for this view, however. In so far as the claim of presentness is tied to the thesis about imagined perceiving, it is threatened by Currie's suggestion that some fictions essentially involve *unobserved* events, and so are not appropriately engaged with by imagining ourselves to be observing them (Currie 1991; 1998: 277–8). Our engagement with these fictions, at least, has to be, as he puts it, 'impersonal'. Another difficulty, also raised by Currie, is presented by *anachrony*, where a story jumps forwards or backwards in time, so that the order of fictional events is not the same as the order of our engagement with them. If we privilege a certain scene in the fiction as present, then it should follow that a flashback is of the past, but this is inconsistent with the claim of presentness, which entails that when we are actually watching the flashback, it is (fictionally) present (Currie 1992).

In the face of these difficulties we have three choices:

(i) Distance the idea of a fictional A-series from ideas such as the claim of presentness, self-location in the fiction, and imagined observing. There is a fictional A-series in the sense that it is part of the *content of the fiction* that a particular event is past, or present, or future. The fictional A-series position of an event is independent of the actual A-series, position of our engaging with it. There is, then, within the fiction itself, some objective determinant of the fictional present.

(ii) Continue to base the fictional A-series on our imaginative self-projection into the fiction, but concede (a) that there may be some fictions for which such self-projection is inappropriate, and therefore that we should not construct a fictional A-series for such fictions (though this is clearly not an option for the defenders of the definability thesis); (b) that in anachronous fictions, what is fictionally present does not stay constant, but hops about from later to earlier times.

(iii) Reject all talk of a fictional A-series, but view fictional time in B-series terms only.

In §8.4, we shall be concerned amongst other things with the fortunes of (i), and in §8.5, with those of (ii) and (iii).

8.3 FUTURE CONTINGENTS IN LIFE AND FICTION

We have looked at the FFF thesis's assumption of a fictional A-series. Let us turn now specifically to the idea of fixity. Now, whether or not certain fictional futures are fixed, there is widespread support for the non-fixity of *our* future. Aristotle offers the failure of bivalence for future-tensed statements concerning contingent matters of fact (future contingents, in short) as one possible escape route from fatalism (Ackrill 1963: 9.19^a). We may find the fatalist argument for an unfixed future—that if what will be, will be there is nothing we can do about it—unconvincing, and yet, for quite different reasons, find the idea of an unfixed, indeterminate future compelling. For one very natural explanation of the direction of time is in terms of the following asymmetry between past and future: whereas any past- and present-tensed statement has a determinate truth-value, no non-necessary statement about the future has such a value. As we might put it, there are facts of the matter as to what *has* happened, but no facts of the matter as to what *will* happen: the past is ontologically determinate, the future ontologically open. Trivial future-tensed statements, such as 'It will be Wednesday tomorrow', uttered on Tuesday, or 'No man will ever be a bachelor during his marriage', are of course true, but not by virtue of any future state of affairs. The non-fixity of the future relates specifically to future *contingents*: it is these that lack a truth-value. (I shall ignore here, as being irrelevant to the discussion of fictional time, the consideration that, given a deterministic universe, logically contingent statements about the future may nevertheless be *physically* necessarily, and so have a truth-value.)

Suppose, then, that the future is in fact unfixed. Is this a reason to infer that the future must also be unfixed *in fiction*? This is not an inference that we should rush to make, since, in general, what is actual places very few constraints on what can be fictional. However, there is a difficulty here, which we might attempt to articulate as follows. To the extent that a fiction represents later states of affairs as determinate, it is not representing the future at all. For if the future is in actual fact unfixed, this is not merely an accidental feature of the future but an essential and defining feature of it. Yet, if the FFF thesis is correct, what some fictions represent under the guise of the future *is* fixed, at least with respect to a certain range of states. Therefore, whatever those fictions are representing, it is not the future. But this conflicts with the suggestion that we can intelligibly raise questions about what *will* happen in a fiction.

It is, however, a commonplace that fictions may take liberties with facts concerning actual events, places, or people. Robert Bolt's *A Man for All Seasons* puts into the mouth of Sir Thomas More words that he never uttered (as well as some that he did). Historical evidence suggests that Shakespeare misrepresents the character and deeds of Richard III. If a fiction can alter an object's contingent attributes, without that undermining the capacity to represent that object, why should it not alter an object's *essential* attributes? Let us say, plausibly enough, that being human is an essential property of yours. There is surely nothing to stop me devising a fiction in which you (yes, you yourself) appear as a gigantic beetle, as Kafka did with the unfortunate—although purely fictional—Gregor Samsa. If you can appear in a fiction bereft of some essential property, why should the future not similarly appear in fictions bereft of its essential indeterminacy?

There is, however, this difference. When I construct a fiction about you, I have some existent object I can first identify and then put into a fictional context. Since I have already identified you, there is no further problem over how you can be identified in the fiction. But if the future is unfixed, then there is nothing to identify in this way. The future is, on this view, essentially unreal. So how do I refer to it, in order to make it the subject of a fiction?

All this, arguably, is simply misguided. When we say 'The future is unfixed', we are not attempting to refer to some strange object whose nature is entirely indeterminate. We simply mean that future-tensed statements have no truth-value. And there seems, on the face of it, no problem at all in saying that, outside the context of fiction, future-tensed statements lack determinate truth-values, but within that context, some of them can have determinate truth-values. There can be no objection, then, from defenders of the actual non-fixity of the future, if we make it a constraint on an account of fictional truth that it be consistent with their position. What could be more reasonable? In fact, this seems so harmless and unexceptionable a constraint (if, that is, you happen to believe that the future is unfixed), that it seems hardly worthwhile making it.

Well, we shall see.

8.4 ACCOUNTS OF FICTIONAL TRUTH

Our task now is to consider how the FFF thesis can be integrated into accounts of fictional truth, bearing in mind the constraints imposed by the previous two sections: we have to be able to identify what within the fiction determines A-series position, and we have to allow that the actual future is unfixed.

8.4.1 Possible worlds

Consider first an account offered by David Lewis:

A sentence of the form 'In the fiction f, ϕ' is non-vacuously true iff some world where f is told as known fact and ϕ is true differs less from our actual world, on balance, than does any world where f is told as known fact and ϕ is not true. It is vacuously true iff there are no possible worlds where f is told as known fact. (D. Lewis 1978: 270)

For the sake of stylistic consistency with later accounts that we shall consider, let us (ignoring the issue about vacuity) paraphrase this as follows:

(L1) It is true in fiction F that p if and only if, within the class of possible worlds where F is told as known fact, there is at least one world where p is true and which is closer to the actual world than any world in that same class where p is not true.

But if it is a necessary non-fictional truth that the future is unfixed, then in no worlds will the future be fixed, so Lewis's account will deliver the result that fictional future contingents lack a truth-value, conflicting with the FFF thesis.

(L1), the doctrine of the essential unfixity of the future, and the FFF thesis thus form an inconsistent triad. To put it another way, (L1) entails that, given the actual unfixity of the future, fictions such as *Macbeth, Time and the Conways*, and *The Time Machine* do not, appearances aside, contain representations of a fixed future.

In Lewis's second account of fictional truth, 'the actual world' in the analysis is replaced by the set of 'belief worlds', these being the possible worlds where what is overtly believed in the author's community is true (or largely true):

A sentence of the form 'In the fiction f, ϕ' is non-vacuously true iff, whenever w is one of the collective belief worlds of the community of origin of f, then some world where f is told as known fact and ϕ is true differs less from the world w, on balance, than does any world where f is told as known fact and ϕ is not true. It is vacuously true iff there are no possible worlds where f is told as known fact. (D. Lewis 1978: 273)

Paraphrased:

(L2) It is true in fiction F that p if and only if, within the class of possible worlds where F is told as known fact, there is at least one world where p is true and which is closer to any given belief world than any world in that same class where p is not true.

Now it may be that one of the overt beliefs in the author's community is that the future is fixed. Nevertheless, *if* the hypothesis that the future is unfixed is necessarily true, then there are no possible worlds in which the future is fixed (unless, unlike Lewis, we have an intensional conception of possible worlds—if e.g. we consider them as merely epistemically possible worlds). *A fortiori*, there would be no belief worlds—as Lewis understands the term—in which the future is fixed. This is an instance of a general difficulty with both of Lewis's accounts: since truth in fiction is for Lewis truth in possible worlds, how can he accommodate inconsistent fictions? One strategy that Lewis suggests is to decompose the fiction into consistent fragments (1983*b*: 277–8), but it is not clear that the denial of some metaphysical truth can be so decomposed: the inconsistency here seems irreducible to atomic propositions that generate an inconsistency only when combined.

If we are prepared to regard the unfixity of the future as merely a contingent matter, then these worries evaporate. For then there *will* be worlds in which the future is fixed. This, though, is not something that is likely to appeal to defenders of unfixity, for metaphysical theses, arguably, are, if true, necessarily so.

Neither of Lewis's accounts, however, offers a way of meeting our first constraint: namely, that there be something within the fiction to determine A-series position. This is not to say that (L1) and (L2) are actually inconsistent with fictional future truth, just that they provide no mechanism for determining what counts as *future* in the fiction.

8.4.2 The fictional narrator

Such a mechanism is provided by the third account of fictional truth I want to consider: namely, that provided by Gregory Currie. In place of Lewis's possible worlds, Currie substitutes sets of beliefs—the beliefs of a *fictional author*. The fictional author is a character within the fiction itself, who narrates the fiction as truth, and who is not necessarily to be identified with the explicit narrator of the fiction. Since the word 'author' in this context tends to suggest someone who precisely is *not* part of the fiction, but is the creator of it, it may be less misleading to talk of the *fictional narrator*. With this change in place, the account is as follows:

(C) It is true in fiction *F* that *p* if and only if it is reasonable for the informed reader to infer that the fictional narrator of *F* believes that *p*. (Currie 1990: 80)

Since it may, for some fictions, be reasonable for the informed reader to infer that the narrator of the fiction believes some metaphysically impossible proposition, there is no obstacle to the idea of some fictions representing the future as fixed. Further, unless we imagine him or her to be a timeless being, the fictional narrator presumably has a temporal perspective on fictional events. It is this perspective that can determine which events are past, which present, and which still future. Moreover, it is a device within the fiction that does so, not something we import into the fiction.

So, a given future-tensed proposition 'It will be the case that *p*' is true if the fictional narrator believes that it will be the case that *p*. But what would justify us in ascribing such a belief? Unless we ascribe to him the capacity for foreknowledge, we cannot be justified in inferring that he has any particular belief about the future. And it would not be reasonable to assume that the fictional narrator had knowledge of the future—not, at any rate, unless the text gave one some fairly explicit evidence that such knowledge was readily available in the world of the fiction. Of course, *Macbeth* does provide evidence of foreknowledge (perhaps one of the witches is the fictional narrator!), and indeed it supports the FFF thesis precisely because there is evidence that some of the characters have knowledge of the future. But not all fictions that support the FFF thesis provide such evidence. *Time and the Conways*, for example, does not. No one in that play can see the future (which is just as well, considering), and although the hero of *The Time Machine* does come to have knowledge of the future, it is not *qua* future but *qua* present that it reveals itself to him. The problem, then, is that, whatever we say about the fictional narrator's perspective, we cannot, in the absence of evidence of foreknowledge, ascribe to him beliefs about the future. And since (C) links the author's beliefs with fictional truth, we cannot (again without recourse to foreknowledge) account for the fictional truth of future-tensed statements in e.g. *The Time Machine*.

There is yet another alternative, and that is to abandon the idea that the fictional narrator has a temporal perspective at all. In so far as he has a perspective, it is the timeless perspective of a deity. But it is hard to see how this would help us. For then he would have no tensed beliefs at all, and so, *a fortiori*, no future-tensed beliefs.

It seems then, that although the device of the fictional narrator can be viewed as something within the fiction that determines the fictional A-series, it cannot be used to determine the truth of future-tensed propositions. Conversely, in so far as it can used to determine all fictional truths, it cannot provide an A-series.

8.4.3 Imagination

In a later paper, Currie proposes to show how tensed fictional truths can, at least in principle, be generated from a thesis which links fictional truth with imagination. The thesis in question is this:

Basic Thesis of Fiction (BTF): What is fictional in F is just what F makes it appropriate for the reader of F to imagine. (Currie 1998: 270)

(Currie takes 'It is fictional in F that p' as equivalent to 'It is true in F that p'; there is no suggestion of nested fictional operators, making something *fictionally* fictional.) The phrase 'is just' indicates that what we have here is a biconditional. To generate tensed truths in fiction from this thesis, as Currie remarks in a footnote (1998: 270 n. 3), we need a time-indexed version of it, perhaps along the lines of the following:

Time-indexed BTF: For any fictional time t, it is true in F at t that p if and only if F makes it appropriate for the reader to imagine that p is the case at t.

So when Chorus at the beginning of Act IV in *Henry V* utters the words

> Now entertain conjecture of a time
> When creeping murmur and the poring dark
> Fills the wide vessel of the universe.
> From camp to camp, through the foul womb of night,
> The hum of either army stilly sounds . . .
> . . . and from the tents
> The armourers, accomplishing the knights,
> With busy hammers closing rivets up,
> Give dreadful note of preparation.
> The country cocks do crow, the clocks do toll,
> And the third hour of drowsy morning name

we are invited to imagine that it is the night before Agincourt. That is, we are (apparently) invited to imagine a *present*-tensed truth. And perhaps also a future-tensed one, for we know that when the sun rises, the battle will begin. (The character of Chorus is actually a rather anomalous one. He is not the fictional narrator, for, although he shares a stage with Pistol and the rest, Chorus is located in our world, not the fictional one, and exhorts us to use our imaginations to transmute the representations on the stage into real events.)

A brief aside before we proceed further. Currie does not intend the BTF to replace (C). Indeed, the BTF seems incomplete as a theory, at least in epistemological terms. It needs to be supplemented by something that tells us how we work out what the fiction makes it appropriate to imagine, and this role can be occupied by the earlier account. Of course, this very dependence of the BTF on a more detailed account is problematic, for if that account is at odds with the FFF thesis, conjoining it with the BTF will not help.

Look again at the time-indexed BTF. The point of time-indexing is that, if p is a tensed truth, there will be times during our engagement with the fiction when it is appropriate to imagine that p, and times when it is not appropriate. Thus, a present-tensed proposition may be imagined as true at one time, but at later times the corresponding *past*-tensed truth must be substituted. Now, when we say that 'F makes it appropriate for the reader to imagine that p is the case at t', do we mean the fiction as a whole, from beginning to end, makes it appropriate to imagine that p, or that precisely at t the fiction does so? And should we interpret 'imagine that p is the case at t' to mean 'imagine (at whatever time) that p is the case at t', or 'imagine at t that p is the case?' Since the two issues are related, we have two possible readings of the time-indexed BTF:

(a) For any fictional time t, it is true in F at t that p if and only if F as a whole makes it appropriate for the reader to imagine that p is the case at t.

(b) For any fictional time t, it is true in F at t that p if and only if F at t makes it appropriate for the reader to imagine at t that p is the case. (Cf. Currie 1998: 270 n. 3)

Here we seem to be faced with a dilemma. Suppose we opt for (a). Then *every* complete and temporally structured fiction will perforce represent the future as fixed, by virtue of the fact that the whole fiction will determine the truth-value of at least some statements of the form 'It is the case at t that it *will* be the case that p'. But surely, if the future is in fact unfixed, fiction cannot be prohibited from representing it as unfixed! Suppose instead we opt for (b). This formulation seems far too restrictive, for it may be only at a later fictional time that the reader is in a position to imagine, as opposed to consider hypothetically, what is the case at t. Ian Pears's *An Instance of the Fingerpost* consists of four accounts, by different narrators, of certain events in seventeenth-century Oxford. The narratives put quite different constructions on these events, so that one simply has to remain agnostic, for a large part of the book, as to what (if anything) is the fictional truth of the matter. Only towards the end (and perhaps not even then) can we be more confident in our imaginings. So if we accept (b) without modification, then it seems that we have to settle for a large amount of indeterminacy concerning fictional pasts, as well as futures, but we would not on that account want to say that such fictions thereby represent the *past* as unfixed. On the other hand, if we modify (b) so as to take into account the idea that later imaginings fix the truth of earlier times, it simply collapses into (a), and we have lost the possibility of an open fictional future.

8.4.4 Make-believe

We have looked at a variety of devices to account for fictional truth: states of affairs in other possible worlds, the beliefs of a fictional narrator, our own imagination. The list is not exhaustive, however, and before presenting a proposal for accommodating the FFF thesis to an account of truth in fiction, we should consider one more analysis: namely, Alex Byrne's.

Byrne rejects Currie's account (C) on a number of grounds: first, it cannot account for 'mindless' fictions, i.e. those that are explicitly not narrated by anybody; second, it is not always clear how to construct the beliefs of the fictional author; third, since it is reasonable to infer that the fictional author is capable of having false beliefs, the account generates the wrong set of fictional truths (although given (b), there will be some indeterminacy as to which set it does generate). Byrne's own account goes as follows:

(B) It is true in fiction F that p if and only if the Reader could infer that the Author of F is inviting the Reader to make-believe that p. (Byrne 1993: 33)

The capital letters here signify that the author and reader in question are not (typically) the actual author and reader. The Reader is an idealized, particularly well-informed (but not omniscient) reader. The Author is constructed by the Reader, such that the intentions of the Author precisely coincide with what the Reader takes to be those intentions.

The connection between (B) and the BTF is a very close one. Indeed, since 'make-believe that p' could be taken to be equivalent to 'imagine that p', and inferences concerning the Author's intentions provide one means of deciding what it is appropriate to imagine, Byrne's account could be presented as a more specific statement of the BTF. Unlike the BTF, however, (B) is structurally parallel to a plausible account of assertion for ordinary, non-fictional narrative: a given narrative N asserts that p if and only if the reader could infer that the author of N intends the reader to believe that p.

Pursuing this parallel, which Byrne makes explicit, it is evident that this Gricean account of assertion in non-fictional narrative has no difficulty in accommodating future-tensed assertions. Let us say that we are reading a report on global warming. The facts presented concern past and present states of affairs. But we are left in no doubt that the author of the report intends us to form the belief that sea levels *will rise* dramatically over the next decade. Now suppose that the report is a fictional one. We can infer that in the world of the fictional report, sea levels will rise over the next decade. Or can we? We may justly suppose that it is *probable* that sea levels will rise, but since the fiction ends before they actually do so, there is nothing to rule out the following possibilities: (i) governments unite in taking steps to reduce global warming; (ii) the greenhouse effect is countered by the dawning of another ice age; (iii) the Earth is destroyed through collision with a giant meteorite; (iv) time itself comes to an end, etc., etc., all of which would prevent the sea levels from rising. Generalizing, we can say that what happens after the end of some fictional narrative will necessarily be indeterminate.

What marks the difference between the above case and those we were considering earlier—*Macbeth, Oedipus, Time and the Conways, The Time Machine*—is that those fictions (eventually) provide the answers to the question, raised at various points in the fictions, of what will happen. It is true right from the start that Oedipus will kill Laius because, by the end of the fiction, he has done so *and* there is reason to suppose that, in this particular fictional world, the future is fixed. Is (B) any better placed than its predecessors to explain this? I suggest not.

(B) can certainly accommodate the fictional truth of the general proposition that the future is fixed, even if the actual future is unfixed. A natural interpretation of *The Time Machine* will certainly allow one to infer that the Author (indeed, in this case, the humble author too) intends the Reader to make-believe that the future is fixed. The difficulty arises, as with (C), over the truth of specific future-tensed propositions. Halfway through the story, neither we nor the Reader are in a position to judge whether the Author intends us to make-believe that the Morlocks will be defeated by the Eloi. Yet, arguably, it is already true at that point in the story that they do. By the time we are in a position to make up our minds on the matter, we can no longer say 'The Morlocks *will be* defeated', but have to say instead 'The Morlocks *have* been defeated'.

8.5 FICTIONAL TIME AS A B-SERIES

Recall the choices we listed at the end of §8.2:

 (i) Take the fictional A-series as part of the intrinsic content of the fiction.
 (ii) Base the fictional A-series on our imaginative self-projection into the fiction.
(iii) Reject all talk of a fictional A-series, and view fictional time in B-series terms only.

Having looked at a variety of different approaches to fictional truth, we could not find a satisfactory way of combining (i) with the FFF thesis. In so far as we wish to preserve the FFF thesis, then, I suggest abandoning (i). My proposal is that we adopt a combination of the apparently exclusive (ii) and (iii).

The combination of these requires a distinction between the intrinsic content of the fiction, which I shall call the *fictional facts*, and true representations of that content, which I shall call *fictional truths*. This might look like a suspect distinction, but we should reflect that the B-theory of time distinguishes between true A-series representations of reality ('Florence has just played the *Goldberg Variations* on the mouth-organ') and the B-series facts that make them true (Florence plays the *Goldberg Variations* on the mouth-organ at 2300 hrs GMT on 31 December 2005). I am proposing to import this distinction into fiction.

The fictional facts—those that define the fiction itself, and which are independent of any particular engagement with it—should be conceived of entirely in B-series terms, I suggest. So, for example, Macbeth's first encounter with the witches is simultaneous with Banquo's; the knocking at the gate occurs simultaneously with the Porter's soliloquy; the discovery of Duncan's body is earlier than Malcolm and Donalbain's flight from Scotland; the appearance of Banquo's ghost occurs between the murder of Duncan and the final confrontation between Macbeth and Macduff, etc., etc.

This leaves no room for an ontological distinction between past and future fictional fact. Later events in the fiction are as determinate as earlier events. If there is any indeterminacy (and of course in fiction there is always indeterminacy), it has no systematic connection with the direction of fictional time.

We seem to be further away from the FFF thesis than ever. However, although the fictional facts are only B-series facts, we can still make sense of true A-series representations of those facts. But what is required for those representations is something external to the fiction: namely, our active engagement with it. For in imagination, as we noted in §8.2, we can occupy a perspective within the fiction. This imagined perspective combines with the fictional facts to make true our tensed beliefs about the fiction. What makes true our tensed beliefs about the fiction thus precisely mirrors what, according to the B-theory, makes true our ordinary tensed beliefs about the world. And, just as the B-theory requires a distinction between what it is true to say of the world (e.g. that the train has stopped here) and what facts in the world make it true (e.g. that the train stops at Adelstrop at ten to three on the afternoon of 16 November 1999, and I say 'The train has stopped' at that place and at that time), so we can distinguish between what it is true to say of the fiction (e.g. that Banquo is now being murdered) and what facts in the fiction make it true (e.g. that Banquo is murdered at t, and in imagination I locate myself at t). This preserves the FFF thesis, as it permits us to have true future-tensed beliefs about fictional events. It is a (B-series) fictional fact that Macbeth becomes King at fictional time t. Now suppose we are watching the scene in which Lady Macbeth persuades her husband to murder Duncan, and that we are imagining the scene as present. The perspective we imagine ourselves as occupying is one that is prior to t. These two elements—the fictional time of Macbeth's coronation, and our current fictional perspective—combine to make true our belief that Macbeth *will become* King. Because we have driven a wedge between what is fact and what is a true representation, however, we would need to revise the various accounts of fictional truth we have considered by replacing 'It is true in fiction F that p' on the left-hand side of the biconditional by 'The fact that p is one of the facts in fiction F'. The fictional *facts* are purely B-theoretic. In order to accommodate A-series fictional *truths*, we then need to supplement each account with the following schema:

A token belief that event E is present (past, future), entertained at fictional time t, is true in fiction F if and only if it is a fact in F that E occurs at (earlier than, later than) t.

In so far as it is appropriate to form tensed beliefs concerning fictional events, we now have a mechanism for establishing the truth of such beliefs.

A consequence of treating fictional facts as B-theoretic is that all completed fictions are bound to represent the future as fixed. But what if the story makes it clear that the fictional author believes that the future is unfixed? Then the fiction would be inconsistent, representing the future both as fixed *and* as unfixed. But this, I submit, is not a serious threat. First, some fictions just *are* inconsistent (such as those in which a time-traveller changes the past), but they are still intelligible at some level. Second, it is not clear just what it would *be* for a fiction to represent the future as unfixed. What, to put it in the context of Currie's account, would warrant our supposing that the fictional narrator believes that the future is unfixed? A mere statement in the narrative to this effect is not enough, since the informed reader will always be on guard against simplistically identifying everything that is stated in the text as an expression of what the fictional author believes. The fact that some statement made within the fiction is at odds with something we take to be fictional truth is at least *prima facie* evidence that the statement in question is not an expression of the fictional narrator's beliefs.

There is a lingering worry over the compatibility of Byrne's account with our proposal. We may still want to make room for the intuition that, as consumers of fiction, we are intended by the author to form tensed beliefs. Now, on (the revised versions of) Lewis's and Currie's accounts, it does not follow that there are A-series fictional facts, but on Byrne's account it may appear to. But perhaps we can exploit Byrne's distinction between the (real) author and the (constructed) Author—and indeed between the (real) reader and the (ideal) Reader—to get over the difficulty. Let us say that the actual author does intend us to form beliefs as to what is about to happen in the fiction. This will not be sufficient to guarantee the truth of those beliefs. In fictions with surprise endings (detective novels, farces) the author may intend us to form *false* future-tensed beliefs. What matters, as far as the facts of the fiction are concerned, is what the *Author* intends us—or rather, the Reader—to believe. And what the Author intends the Reader to believe, I suggest, is captured entirely in B-series terms.

I should now like to return to an issue briefly touched on in §8.2, that of interactive fictions. Is the difference between interactive and non-interactive fictions precisely that the former, but not the latter, represent the future as unfixed? We are not forced to say this. Truth in interactive fictions may be just as determinate as it is in non-interactive fictions. Consider, for instance, the B-theoretic reconstruction of Currie's fictional author account. Can this be applied to interactive fictions? It all depends on whether we, *qua* part-creators of the fictions with which we are interacting, thereby play the role of the fictional narrator. If we do, then the account is not applicable, as we do occupy a temporal perspective, and many of our beliefs are tensed. But there is reason not to identify ourselves with the fictional narrator in such cases. For, even though we partly determine the course of the fiction, it is still possible for us to be mistaken about the fictional facts. We may misinterpret some fictional situation, or fail to see the fictional consequences of our actions. If this is so, then even in interactive fictions, the viewpoint of the fictional narrator can be a timeless one, and if it is, the fictional facts will form only a B-series. So the gap between interactive and non-interactive fictions closes, in that the former turn out to be as complete as the latter. It closes still further once we realize that, on the B-theoretic account, both kinds of fiction require our interaction with them for the generation of tensed fictional truths.

My appeal to our imaginative projection of ourselves into fiction may well be objected to (as mentioned earlier, Currie has worries about such appeals). But it is important to make clear that I am not committed to the rather strong view that an understanding of fiction, or awareness of the fictional facts, requires such self-location, merely that the formation of *tensed* beliefs about what is fictionally true does so.

Can we draw any conclusions concerning the nature of real time? It would not be appropriate to try to generate a metaphysical result from a discussion of fiction. On the other hand, the fact that the B-theory of time can, suitably adapted, solve a problem in aesthetics is a modestly gratifying result for proponents of the theory. In addition, its success in this context provides further evidence against the suggestion, the truth of which would completely undermine the B-theory, that B-series expressions are parasitic on A-series ones.

And so we leave Lady Macbeth reading her husband's letter, Laius musing on the Oracle's prophecy, the Conways contemplating a bright future, and the time-traveller about to set off for the land of the Morlocks and the Eloi. Their fate is sealed—as, perhaps, is ours.

9

The Unity of Time and Narrative

Another time has other lives to live.
W. H. Auden, 'Another Time'

9.1 THE POSSIBILITY OF DISUNIFIED TIME

Kant held that space and time were unique: there is only one space, and only one time series. Or, in other words, that any objects bearing spatial and temporal relations to *any* other object would bear them to *all* other such objects: 'Different times are but parts of one and the same time' (Kant 1929/1787: A32). The idea of *disunified* time (or space), in contrast, is the idea that there might be objects that bear temporal (or spatial) relations to some other objects, but not to all other objects. Thus, there might be two objects which, though each is in *a* time, are neither simultaneous nor stand in any relations of temporal order to each other.

A number of writers since Kant have been prepared to entertain the view that space might not be unique, that there might be universes with their own spaces, so that an object in one universe would bear no spatial relations to any object in another universe. But the belief in the uniqueness of time has proved to be rather more recalcitrant. Anthony Quinton, for instance, having invented a myth demonstrating the conceivability of evidence for disunified space, argues that an analogous myth for time cannot be constructed:

The moral of these unsuccessful attempts to construct a multi-temporal myth is the same in each case. Any event that is memorable by me can be fitted into the single time-sequence of my experience. Any event that is not memorable by me is not an experience of mine . . .

I conclude, then, that we can at least conceive of circumstances in which we should have good reason to say that we knew of real things located in two quite distinct spaces. But we cannot conceive of such a state of affairs in the case of time. Our conception of experience is essentially temporal in a way in which it is not essentially spatial. (Quinton 1962: 219)

That last remark is entirely Kantian. Kant described space as the form of *outer* sense, meaning that experience represented external objects as spatially related to each other. But time he described as the form of *inner* sense, meaning that experiences presented themselves as temporally related. (Kant 1929/1787: B37) So the sense of time order is for Kant essential to all experience, the sense of spatial order essential only to some of it.

But how do we move from the fact, if it is one, that experience presents itself as temporally unified to the unity of time? For Kant, there is no gap, since for him time is nothing more than a form of intuition: it is not a feature of the world as it is in itself. To say that time is necessarily unified is just to say that experience necessarily presents itself as such.

Does experience present itself as temporally unified? Arguably, it does not do so for the young child. Friedman describes the young child's experiences as consisting of 'islands of structure': localized patches of events which are internally temporally structured, but which are not integrated into an all-encompassing framework (Friedman 1990: 93). One might argue that a mature concept of time involves just such a framework, but this seems less a phenomenal aspect of experience than the result of reflection on experience. It is not at all clear that our memories can all be related to each other temporally. They all concern 'the past', certainly, but we are not always able to order them in time. As an extreme case consider the situation of 'split-brain' patients, whose corpus callosum, the bridge of tissue and nerves linking the two hemispheres of the brain, has been severed or temporarily deactivated by freezing. In certain experimental conditions, the hemispheres process quite different sensory inputs, and can communicate independently (Sperry 1964; Nagel 1971). Suppose, then, that a patient had his corpus callosum temporarily frozen, preventing communication between the hemispheres, and was then placed in a situation where one set of information was fed to one hemisphere and another set of information fed to the other. When the corpus callosum was unfrozen, the patient would have memories of x happening and also of y happening, but no sense that x and y stood in any temporal relation to each other.

In any case, as we saw in Chapter 6, temporal features of the world such as duration and precedence can only play the explanatory role

they do if they are mind-independent. What, then, if we distinguish
between the experiential representation of time, on the one hand, and
the external time it represents, on the other? Is there any reason to
suppose that external time must be unified?

Although he is clearly influenced by Kant, Quinton does not explicitly treat time as mind-dependent. If we recall the context in which
he was writing, however (Oxford in the early 1960s), it is entirely
plausible to suppose that he took it as a necessary step in making the
hypothesis of disunified space or time intelligible that one be able to
describe circumstances in which such a hypothesis would have empirical
grounding. There is, in other words, likely to be a verificationist premiss
lurking in the background.

If, then, we are neither inclined to treat time as mind-dependent,
nor in the grip of verificationism, why should we suppose time to
be necessarily unified? The remark by John Lucas quoted earlier in
Chapter 3 suggests another line of argument:

It would be unintelligible for me to offer a frame of temporal reference within
which I could not refer to the date at which I was then speaking. It is part
of the concept of time that it is connected to us, whereas it is not absolutely
necessary . . . that space should be connected to us. The essential egocentricity
of time is reflected in the ineliminability of tenses. (Lucas 1973: 280)

Although Lucas describes it token-reflexively ('the time at which I was
then speaking'), the key idea here is that of the present (and, relatedly,
of the past and future), which Lucas is suggesting is essential to time.
Is there, then, a connection between the reality of the A-series and
the unity of time? It certainly seems initially as if there is, for only if
we are in a time series can we refer to any time within that series as
'present'. It would make no sense to say of a time within a parallel series
that it was present, for the present is where in time *I* am. Ironically,
this thought depends on the kind of token-reflexive view of tensed
terms that we associate with the B-theory. If 'the present time' means
nothing more than 'the time at which this remark is made', it need not
stand for any ontologically privileged position. If, on the other hand,
the present is a privileged moment, its status should not depend on
anyone's perspective. On the contrary, a person's perspective would be
determined by which time was objectively present. This leaves the door
open to a time in some other time series being present, without my
or indeed anyone having *any* temporal perspective on that time. This
seems to be McTaggart's own view:

No doubt in such a case [i.e. that of disunified time], no present would be the present—it would only be the present of a certain aspect of the universe. But then no time would be the time—it would only be the time of a certain aspect of the universe. It would be a real time-series, but I do not see that the present would be less real than the time. . . . if there were any reason to suppose that there were several distinct B-series, there would be no additional difficulty in supposing that there should be a distinct A-series for each B-series. (McTaggart 1927: 30)

Defenders of the A-series could in fact go either way on this issue. If they opted for what we called in Chapter 3 the simple A-theoretic semantics, then the truth-conditions of thoughts like '*x* is present' would be independent of the thinker's position, and so such thoughts could quite meaningfully be applied to other time series. If, on the other hand, they opted for a more complex, date semantics, according to which the truth-conditions of '*x* is present' and the like are sensitive to an A-series context of belief, then they could not be so applied. Nevertheless, this second view would still be compatible with the idea of an A-series unrelated to our own.

The conflict between the idea of time's unity and the A-theory involves more than the mere existence of the A-series; it involves two other ideas typically held by A-theorists, expressed in the following remarks by Arthur Prior:

If, as I would contend, it is only by tensed statements that we can give the cash-value of assertions which purport to be about 'time', the question as to whether there are or could be unconnected time-series is a senseless one. We think we can give it a sense because it is as easy to draw unconnected lines and networks as it is to draw connected ones; but these diagrams cannot represent time, as they cannot be translated into the basic non-figurative temporal language. If we try to translate them, we produce contradictions. . . . like 'There are things going on which neither are going on, nor will be going on, nor have been going on.' (Prior 1967: 198–9)

One of the purposes of Prior's construction of different tense-logical systems was to show how different topologies for time could be represented in tensed (and so A-theoretic) terms. The density of time, for instance, which we might otherwise express in terms of moments—that between any two moments there is always a third—can instead be represented by the axioms Pp → PPp and Fp → FFp (if it was/will be the case that p, then it was/will be the case that it was/will be the case that p). How, then, should we represent the idea of time's being disunified? When we introduced the idea above, we put it in terms of two objects that were neither simultaneous nor temporally ordered with respect to one another. But this is to put it in terms of the B-series relations, which, according to the A-theorist, supervene on A-series positions. How do we express disunified time in A-theoretic terms? We could try saying that time is disunified if there are objects that are neither past, present, nor future. But this leads to two problems. One, not noted by Prior in the passage above, is that this could obtain if there existed abstract, timeless objects, such as numbers and universals. But the existence of abstract objects has no implications for time's topology. We therefore have to convey that objects in other time series are *temporal*, and how can we convey that except by saying that they are either past, present, or future? The second problem, which is noted by Prior above, arises from the tensed nature of existential assertion (another characteristic thesis of the A-theory). Suppose, like Prior, we are presentists. Then an assertion like '*x* exists' is taken to be equivalent to '*x* is present'. To say that there existed objects which are neither past, present, nor future would then be to utter the evident self-contradiction that there are present objects that are neither past, present, nor future. Anyone who thinks that '*x* exists' is equivalent to one or more of the disjuncts in '*x* is present or *x* is past or *x* is future' faces a similar problem.

We might just note in passing that there is no direct conflict between presentism and disunified time. Even if '*x* exists' means '*x* is present', there is no formal contradiction in supposing that different, unrelated times are present (provided the presentist opts for the simple A-theoretic semantics). On the other hand, anyone who held this would have to give up the usual presentist assumption that everything that exists is simultaneous. Giving this up would be awkward for the presentist. For if we are willing to contemplate the idea that times unrelated to our present might nevertheless exist, what could be the objection to treating times which were earlier or later than our present as existent?

In summary, if topological assertions about time *and* existential assertions are tensed, in the sense of making implicit and ineliminable reference to the A-series, then we cannot coherently represent the idea of time's disunity.

So much for the case against disunified time. It rests on assumptions that have been among the targets of this book. We can, I have contended, make sense of genuinely tenseless, purely B-theoretic statements. This in itself is not inconsistent with the existence of an A-series. But representing the truth-makers of our A-beliefs in terms of B-facts rather than A-facts makes sense of various aspects of mental representation. I make no apologies, then, for simply assuming, for the purposes of the rest of this chapter, that disunified time is a metaphysical possibility. My concern from now on is with the *fictional* representation of disunified time—or, as we might put it, the disunity of fictional time. The interest of this is threefold. First, if there are fictions that are best interpreted as representing time as disunified, then this puts further pressure on those who deny the coherence of disunified time. I do not put much emphasis on this, however, as those who are convinced by Prior's argument will simply say either that this is not the best interpretation of such fictions or that some fictions are inconsistent. Second, if it turns out that there is an in-built constraint on narrative and fiction such that a single story can only represent a single time series, than we may have found another source of intuitive resistance to the idea of disunified time. Third (a line of thought that runs in the reverse direction), on the principle that anything metaphysically possible is fictionally possible, the metaphysical possibility of disunified time is an objection to those theories of fictionality and fictional truth that require fictional time to be unified.

9.2 DISUNIFIED FICTIONAL TIME

We might say that each fiction has its own time series, in which the events of that fiction are ordered, and although some fictions might be said to share our time series (especially those historical fictions which are based in part on actual events), many do not. So whereas the novels of Dickens are very clearly located in the nineteenth century, it makes no sense to locate *The Lord of the Rings*, for instance, in the Dark Ages, or in any other part of our own history. One might then say that radically different fictional worlds constitute different fictional time series. But that is not the sense of disunified fictional time that I am interested in. A case of disunified fictional time, in my sense, would be one in which a *single* fiction represents more than one time series. Are there such cases?

Consider first those fictions that have other fictions within them. Part of the action of *A Midsummer Night's Dream*, for instance, concerns the performance of a play, an enactment of the story of Pyramus and Thisbe. What we might call the embedded fiction (Pyramus and Thisbe) has its own time series, in which the events of the embedding fiction (the marriage of Hermia to Lysander, for instance, or of Theseus to Hippolyta) have no place. The embedding fiction has another time series. We do not have two independent fictions here, arguably; so is this a case of a single fiction containing two time series? No. From the perspective of the embedding fiction, the embedded fiction is not another time series, but merely the *representation* of a time series. From the perspective of the embedded fiction, the embedding fiction simply does not exist. Of course, there are fictions that break the rules. In Flann O'Brien's *At Swim-Two-Birds*, we find the following passage:

DERMOT TRELLIS, an eccentric author, conceives the project of writing a salutary book on the consequences which follow wrong-doing and creates for the purpose . . .
JOHN FURRISKEY, a depraved character, whose task is to attack women and behave at all times in an indecent manner. By magic he is instructed by Trellis to go one night to Donnybrook where he will by arrangement meet and betray
PEGGY, a domestic servant. He meets her and is much surprised when she confides in him that Trellis has fallen asleep and that her virtue has already been assailed.

Wait a moment. How is it possible for a character in a fiction to inform another character of the activities of the fiction's creator? In one sense, it is not possible. Yet this is indisputably part of what happens in O'Brien's novel. We have here an embedded fiction that leaks, as it were, into the embedding fiction. But we add nothing to our understanding of this extraordinary work by describing it as containing separate time series that fuse at some point. O'Brien is, rather, playing with the relationship between author and character and exploiting our ability to understand the paradoxical and inconsistent. What we are looking for is a fiction that is in all other respects entirely coherent, and requires for its interpretation the assumption of separate fictional time series.

A novel highlighted by Paul Ricoeur as a 'tale about time', and which illustrates a certain kind of disunity, is Virginia Woolf's *Mrs Dalloway*. In this 'stream of consciousness' novel, Woolf, influenced by Bergson's (1910) distinction between historical, external time and psychological time, presents us with the private thoughts and emotions of a number of characters as they cross each other's paths during a single day in London. The narrative shifts from one perspective to another, the sense of a shared, public time provided by the regular striking of Big Ben. Each character, however, has their own psychological time, in which the past can be revisited and events whose physical duration was brief are dilated by emotional significance. Does this provide the kind of example we are looking for? I suggest not. In describing the narrative in terms of parallel psychological time series whose metrics are incommensurable with one another, we might seem to be saying something significant about temporal topology. But I have insisted throughout this essay on a sharp distinction between representation and what is represented. And when we describe *Mrs Dalloway* in more prosaic terms as presenting a number of divergent mental representations of a series of events, the sense of temporal disunity disappears. (Which is not to say that the novel, with its shifting perspectives, does not have anything to teach us about the other issues I have discussed, such as the idea of the fictional narrator or the structure of temporal experience. For Ricoeur's discussion, see his 1985: 101–12.)

Consider next the *Narnia* stories of C. S. Lewis. It is a feature of a child's visit to Narnia that, however long she spends there, hardly any time has passed at home on her return. The elderly professor whose wardrobe first conveys Lucy to Narnia at the beginning of *The Lion, the Witch and the Wardrobe*, suggests to Peter and Susan, who are puzzled by their sister's odd behaviour, that, were there to be another world, it is entirely probable that it would have 'a separate time of its own'. The idea of different times had occurred to Lewis earlier in a theological context. In a series of radio talks entitled *Beyond Personality*, broadcast in 1944, he had suggested that God might have his own time, unrelated to ours. He attempted to make this intelligible to his listeners by means of an (as he conceded, limited) analogy: the time of a novel bears no relation to that of its author, yet the author can intervene at any point in the story to determine what happens (C. S. Lewis, 1952: Part 3).

Despite Lewis's intentions, *The Chronicles of Narnia* do not provide an uncontroversial example of disunified fictional time. The discrepancy between Narnian time and English time is purely metrical. It is still possible to put events in England and events in Narnia into a consistent time sequence: Lucy's first meeting with Mr Tumnus takes place before Peter and Susan's conversation with the professor, which takes place before Aslan's confrontation with the White Witch, etc. The metrical discrepancies between the two worlds could be explained entirely by the different rates of change in each, rather than anything to do with time itself.

The same could be said of Richard Swinburne's (1964–5) fantasy, intended originally to show what might constitute evidence of disunified time. To bring to an end a territorial dispute between two tribes, the Okku and the Bokku, a magician casts a spell, bringing it about, from the Okku's point of view, that the Bokku disappear for twenty years, leaving the Okku in sole possession of the territory for that time. From the Bokku's perspective, however, it is the Okku who disappear, leaving them in sole possession for *thirty* years. What has the magician done? The conclusion we are invited to reach (although Swinburne later retracted the suggestion (1968)) is that the two tribes occupied different and unrelated time streams during the period of separation. (From the tribes' point of view, it would have been better to have made the separation permanent, but then the point of the story would have been lost.) Although the metrical discrepancy makes it impossible to establish simultaneity relations between Okku events and Bokku events during the separation, there are at least consistent accounts that one could tell of the ordering of those events.

Appreciating the need for more radical discrepancies, William Newton-Smith tells a story in which it is impossible to construct a consistent account of the ordering of events in different worlds (1980: 83–95). Unfortunately, the 'disunified time' interpretation of the story is competing with a 'time travel' interpretation, according to which shifting between worlds also involves travelling forwards or backwards through time in unpredictable ways.

A rather different, and much more elaborate, kind of narrative is provided in Fred Hoyle's novel *October the First is Too Late*, a story evidently inspired by quantum theory. One day in September 1966, Britain loses all radio contact with America and Europe. A massive nuclear attack is suspected, until air reconnaissance reveals no physical damage to the American mainland, but something much stranger: it appears as it would have done before 1800. Then a number of soldiers in First World War uniform are picked up from the English Channel. It gradually dawns on everyone that different parts of the world are in radically different time zones: it is 1966 in Britain, 1917 in France, and 1750 in America. This is not a case of global time-travel, as there is no record of this peculiar dislocation having already happened, as there would have been had France in 1917 been aware of it. What, then, has happened? The interpretation that most neatly explains the various incidents of the novel involves several parallel universes, which interact at one point and then branch off again. The France that is still at war is not the France of 'this' world, but a duplicate, following the same history.

Is it essential that we identify stories about different time series in the literature? Could we not simply invent a minimal narrative, which begins: 'There is, not one, but several universes, each with its own space and time. What happens in one universe bears no spatial or temporal relations to anything that happens in any of the others'? Those other fictions, however, have something which that bald passage conspicuously lacks: not just colour and human interest, but narrative coherence. They *tell a story*. What I want to consider now is whether there are essential features of narrative fiction that are at odds with the representation of fictional time. Must fictional time, at least, be unified?

9.3 NARRATIVE COHERENCE AND THE UNITY OF FICTIONAL WORLDS

When E. M. Forster was invited to give the 1927 Clark Lectures in Cambridge, the latest in a series of lectures in English Literature established some forty years earlier, he chose the title *Aspects of the Novel*. He selected seven such aspects: story, people, plot, fantasy, prophecy, pattern, and rhythm. In his lecture on story, he gave his famous characterization of a novel as a book that 'tells a story' and went on to define a story in terms of a single time line:

In a novel there is always a clock. The author may dislike his clock. Emily Brontë in *Wuthering Heights* tried to hide hers. Sterne, in *Tristram Shandy*, turned his upside down. Marcel Proust, still more ingenious, kept altering the hands, so that his hero was at the same period entertaining a mistress to supper and playing ball with his nurse in the park. All these devices are legitimate, but none of them contravene our thesis: the basis of a novel is a story, and a story is a narrative of events arranged in time-sequence. (Forster 1927: 43–4)

It follows that no story could be about two unconnected time streams: anything that presented itself as such a story could only be two stories, with no connection at all, arbitrarily bundled together. Forster, however, is more concerned with those novels that have no time line at all, thus, as he sees it, dissolving into incoherence. He considers Gertrude Stein's attempts in her later work to have 'smashed up and pulverised her clock', i.e. to have abolished any chronology, and concludes:

The experiment is doomed to failure. The time-sequence cannot be destroyed without carrying in its ruin all that should have taken its place; the novel that would express values only becomes unintelligible and therefore valueless. (Forster 1927: 53)

Forster is talking here about a very basic aspect of a story. It is not to do with higher aesthetic properties. As he points out, the notion of *plot* is quite distinct from that of a story, and is 'an organism of a higher type'.

We might take issue with Forster's definition of a story, or, more reasonably, question whether all fictions, even narrative fictions (i.e. those that are about happenings), have to be stories in his sense. But, in defence of Forster, we can identify two reasons to suppose that the unity of a fiction requires a single time line. The first is to do with causality. A coherent narrative fiction is one in which the various characters and events stand in causal relations to each other. They need not be interacting all the time. Indeed, there may be parallel sequences of events that for significant periods in the story are causally isolated from each other. But at some stage they will connect up. The novels of Dickens provide good examples of this. Take the plots of *Our Mutual Friend*. There is, on the one hand, the appearance of John Harmon, his introduction to the Boffin household, and his relationship with Bella Wilfer. On the other hand, we have the parallel plot involving the courtship of Lizzie Hexham by the idle barrister Eugene Wrayburn and the unhinged schoolmaster Bradley Headstone. Very little directly links these plots. But they have points in common: it is Lizzie's father Gaffer Hexham who first discovers John Harmon's papers on a corpse fished out of the Thames. And Mr Boffin at one point consults the lawyer Mortimer Lightwood, a friend of Wrayburn's. This is enough to establish a common time. If these plots had remained completely causally isolated, with no character or circumstance in common at any point, what reason would we have for supposing them to be part of the same fiction? What narrative purpose would be served in juxtaposing them? We are not normally inclined to take a collection of short stories, each of which makes no reference to any other, as parts of a single fictional world. But if all threads of a fiction are, at some point at least, causally connected, then they are also temporally connected.

Newton-Smith's fantasy, introduced in the previous section, might appear to be a counterexample to the idea that fictional causal connections imply fictional temporal connections. In this story, intended to make sense of the idea of evidence for disunified time, the characters are constantly hopping from one time series to the other. This movement of the characters introduces a causal connection between the time series (a resolve made in one series may have consequences for actions in another). This very aspect of the fantasy, however, undermines Newton-Smith's proposal that the experiences of its inhabitants gives them reason for supposing there to exist more than one time series. In the standard narrative, the passage of time is indicated by causal connections: changes in the same character, or events at one point influencing events in the other. This same connection is required in Newton-Smith's fantasy: each time series has its own causal structure—that is what makes it the same time series. We are then supposed to ignore this implication of causality when contemplating world-hopping. But how can fictional causal connections both have, and lack, temporal implications in one and the same fiction?

Hoyle's novel, however, shows how different time series can be combined with a causally coherent narrative. If, rather than characters moving between isolated time streams, the streams themselves converge at some point, then events that belong to different streams can be part of the same fiction by virtue of having common effects, at the point of intersection. Of course, the fusing of two separate time streams is bound to have dramatic consequences, and the anomalies that are apparent at the point of intersection, and which occur near the beginning of *October the First is Too Late*, indicate the previous existence of different streams. Once the past existence of such streams is allowed, it is then natural to imagine different regions branching off into separate time streams again.

A second reason to suppose that fictional unity requires temporal unity has to do with a device discussed in the previous chapter: the fictional narrator. The point of a fictional narrator is to explain those fictional truths that are not part of the explicit content of a fiction, or which seem at odds with what is explicitly presented. What is true in the fiction is just what the fictional narrator believes, which may diverge from what is explicitly narrated. The fictional narrator, recall, is intended to be a character within the fiction, not an idealized version of the author. The advantage of this is that it allows us to read into the fiction a *perspective*—spatial, temporal, or epistemic—for that perspective will be the perspective of the fictional narrator. Now suppose we combine these ideas: first, that the appropriate temporal perspective on the fictional events (i.e. their being viewed as past, present, or future) is that of the fictional narrator, and second, that we are obliged to view the fiction from a temporal perspective (this second idea perhaps arising from the thought, expressed by those remarks by Prior and Lucas quoted in §9.1, that we cannot view time as anything other than an A-series). The conjunction of these ideas implies that there can only be one time series within a fiction: namely, the time series that the fictional narrator occupies. The narrator cannot be in more than one series, any more than we can, and if the narrator were to speak of another, parallel series, it would have to be from *no* temporal perspective.

So either the notion of fictionally disunified time is incoherent, or the device of the fictional narrator is one we should discard, or temporal perspective is not essential to fiction. I am not able to offer a conclusive argument for one of these conclusions, but I would like to offer the following line of thought.

What is fictionally possible we should expect to be a broader, rather than a narrower, category than the metaphysically possible. Unless something involves a clear contradiction, then we ought to be able to construct a fiction about it. Arguably, in fact, metaphysical hypotheses *are* fictions. I do not mean by this to suggest an instrumentalist approach to metaphysics, a view of metaphysical hypotheses which takes them to lack any truth-value. I mean, rather, that such hypotheses are stories of how the world may be: representations whose content and implications we can explore, just as we can with fictions. So the metaphysically possible is also the fictionally possible. The reverse may not hold. Perhaps we can in fiction explore situations that we might recognize as not metaphysically possible (discontinuous existence, for instance), but we need not take a view on that here. Now, disunified time, I have suggested, is metaphysically possible. The one valid argument that has been offered against it, namely Prior's, depends on assumptions I have argued against elsewhere in the book. So we should expect to make sense of the idea of a *fictional* representation of time as disunified.

But now we face a conflict with the plausible notion that temporal unity is a condition for narrative unity. That is, what makes two events, or characters, part of the *same fiction* is that they are represented, directly or indirectly, as being in a single time series. This is typically indicated in fiction by the causal connections linking each fictional event. This is certainly a natural assumption to make of any narrative fiction, and it may explain, in part, our pre-theoretic reluctance to embrace the idea of disunified time. How do we tell *a story* about separate time streams?

There are two ways out of this quandary. The first is to distinguish narrative from non-narrative fictions (or narrative from non-narrative parts of a fiction). A narrative fiction is a story, as Forster defines it: namely, an account of events in a particular time sequence. Non-narrative fictions are non-temporal, describing objects or events in a way that is entirely neutral as to their position in time (A-series or B-series). This second kind of fiction is unfamiliar in a literary context, but entirely familiar in a scientific one. The part of a fiction that describes, or entails, the disunity of time we could therefore describe as non-narrative. The problem with this resolution of the conflict is that, in literary fictions at least, it may be difficult or impossible to separate the narrative from the non-narrative elements. The second resolution is to point to those fictions that represent branching pasts: that is, separate time streams that fuse at some point, as in Hoyle's novel. This permits both the representation of time that is, at some point, disunified, with the causal connections necessary for fictional unity.

In §9.1 we examined the suggestion that disunified time is incoherent because we cannot have a temporal perspective on events in other time streams: we cannot, that is, view them as past, present, or future. We find a similar issue in fiction: if we are obliged to adopt a single perspective on fictional events—that of the fictional narrator, who is limited to a single time stream—then we cannot make sense of disunified fictional time. But, I have suggested, we *can* make sense of disunified fictional time. So we should run a *modus tollens* rather than a *modus ponens* inference: we are *not* obliged to adopt a single perspective on fictional events. This does not mean that it is illegitimate to adopt a temporal perspective on fictional events: there will be circumstances in which it is appropriate, and others in which it is not appropriate. And here we can bring in one of the conclusions of the previous chapter: we do not need a fictional narrator to provide that sense of perspective. Fictional time, like real time, is simply a B-series. Fictional events are not in themselves past, present, or future. We generate an A-series by choosing to adopt the view from a particular location within that fiction.

Does this make the idea of the fictional narrator *de trop*? Not necessarily: there may be other roles for this device to play. But one of the supposed advantages of that device, that it provides a fixed temporal perspective on fictional events, turns out to be no advantage at all.

Conclusion

> 'You see, my dear Watson'—he propped his test-tube in the rack and began to lecture with the air of a professor addressing his class—'it is not really difficult to construct a series of inferences, each dependent upon its predecessor and each simple in itself. If, after doing so, one simply knocks out all the central inferences and presents one's audience with the starting-point and the conclusion, one may produce a startling, though possibly a meretricious, effect.'
>
> Arthur Conan Doyle, *The Dancing Men*

Two episodes in the history of science provide an analogy for the two parallel lines of investigation in this project.

At the beginning of the nineteenth century, the physicist, medic, and Egyptologist Thomas Young was studying the nature of light. One of the questions that concerned him was whether light should be conceived of in terms of waves or particles. Another was how it is possible for us to perceive the range of colours that we do. Do we have separate receptors for each colour? Given the huge number of distinguishable colours, this did not seem very likely. Young noticed that light of any hue could be producing by mixing different combinations of just three colours: green, red, and violet. (Green and red light, for instance, would produce yellow light; green and violet would produce blue; all three would produce white light.) On the basis of this, he proposed that our perception of colour depends on just three kinds of receptor, one for each of these 'principal' colours.

Over a century later, in 1912 and 1913, William Bragg and his son Lawrence were conducting experiments with X-rays at Leeds, where William was Cavendish Professor of Physics. It had already been discovered by Walter Friedrich and Paul Knipping, acting on a suggestion by Max Von Laue, that if X-rays were passed through a crystal and then allowed to impinge on a photographic plate, the resulting photograph showed a regular pattern of spots, as a result of the rays being scattered by the regular arrangement of atoms in the crystal. The Braggs used a variant of this method (involving an electroscope attached to a chamber containing a gas which became ionized and so able to conduct electricity when exposed to X-rays) to determine how the atoms in a crystal of sodium chloride were arranged. The method was subsequently employed on more complex, organic molecules, and led ultimately to the discovery of the structure of DNA in 1953.

The question Young had asked himself could be formulated in this abstract form: What mechanism of representation could explain how *those* aspects of the external world (different wavelengths of light) could lead to *these* representations (perceptual experiences of colour)? Similarly, but also contrastingly, the question the Braggs had posed could be formulated: 'What feature of the external world (the relative position of atoms in the crystal lattice) could explain how *that* representational mechanism (the diffraction of X-rays by atoms) could lead to *this* representation (the pattern of ionization intenities)?' In both cases, we have three components of an equation: representation, representational mechanism, and item represented. In each case, the problem was to solve for one of the components given information about the others. Our investigation has pursued questions of a similar structure: what mechanisms of representation, in its various forms, can explain how *those* aspects of time can be represented in *these* ways? What aspects would time have to have to explain how *those* mechanisms of representation give rise to *these* mental representations? In our case, all three components of the equation turned out to be problematic. We have well-established theories of memory, perception, art, and fiction, but are they adequate when *time* is the object of representation? There are the aspects we commonly attribute to time: passage, order, duration, and unity. But are these features merely mind-dependent? There were even doubts about the intrinsic features of time representation, as we saw in the case of the phenomenology of succession. Nevertheless, by filling in accounts of two of the components, we could sensibly ask about the nature of the third. The result is, I hope, a coherent picture of time itself, the mechanisms by which we represent it, and the features of the consequent representations.

To summarize the main features of that picture very briefly:

1. We can make sense of the idea of an 'objective' representation of time, one that is from no temporal perspective; or, to put it in other

words, the idea of a pure B-series (involving just simultaneity and precedence) is coherent, and not parasitic on the idea of an A-series (past, present, and future).

2. The epistemic link between episodic memory and the original experience which gives rise to it is best understood in terms of a B-theoretic conception of the truth-makers of our beliefs—time, that is, does not pass.

3. There is no compelling argument, despite its intuitive appeal, from the features of our temporal experience to the A-theory of time.

4. The causal theory of perceptual knowledge should be amended to include a wider relationship between truth-maker and knowledge, namely an explanatory one.

5. As a consequence, duration and precedence can be seen as entirely mind-independent features of the world, despite the fact that our perception of them is indirect.

6. Static images such as paintings can, given a plausible theory of depiction, be said to depict both change and the static components of that change.

7. The best interpretation of fictional truth treats it as tenseless—that is, as representing fictional events as a pure B-series; we can make sense of the idea of temporal perspective within fiction, but:

8. That is not best interpreted via the device of the fictional narrator, nor is it essential to fiction, as the possibility of disunified fictional time shows.

Of course, all this is controversial. I certainly hope so, for an uncontroversial discussion would also be an uninteresting one. But for those who disagree with my conclusions, I offer the methods employed here as a way of casting light on time and the mind's engagement with it, methods which themselves are a promising source of controversy.

I also hope that this study indicates the possibility of a *rapprochement*—or perhaps *approchement* would be a better word—between two different traditions: on the one hand, the analytic philosophy of time, which has long been dominated by the philosophy of space-time physics and the semantics of tense, and on the other, the phenomenological approach associated with writers such as Bergson and Husserl, which on the whole has been sceptical of metaphysical inquiry. I have said very little about that phenomenological tradition, but the subject of

Chapter 5 in particular—namely, the argument from experience—links issues in the phenomenology of time with those concerning the nature of time itself. Phenomenology and metaphysics need not be strangers.

Other partnerships in this area are likely to bear fruit. Ernst Gombrich's brilliant fusion of aesthetics with the psychology of perception in his various writings showed what could be achieved by bringing together different disciplines. One could not claim a comparable interdisciplinarity for this short essay on temporal representation; but if its methods and conclusions have any merit, philosophers of time may have as much to learn from psychology as from physics. For unless we study the mind's methods of representing the world, there is something whose truth and significance for our understanding of reality we cannot explain: that time, in all its aspects, pervades the whole of our experience, even when the object of experience is experience itself. In this, time has only one rival. 'I have felt', said Wordsworth,

> A presence that disturbs me with the joy
> Of elevated thoughts . . .
> Whose dwelling is the light of setting suns,
> And the round ocean and the living air,
> And the blue sky, and in the mind of man.
>
> 'Lines composed a few miles above
> Tintern Abbey'

Bibliography

I have included in this bibliography a number of useful resources on temporal representation, beside those explicitly mentioned in the text.
Note: Details of first publication are given below. Where details of a reprinted version are also given, page references in the text are to this version.

Ackrill, J. L. (1963), (ed.), *Aristotle's Categories and De Interpretatione*, Oxford: Clarendon Press.

Ayer, A. J. (1940), *The Foundations of Empirical Knowledge*, London: Macmillan.

Baddeley, Alan D. (1976), *The Psychology of Memory*, New York: Harper and Rowe

Barnes, Jonathan (1982), *The Presocratic Philosophers*, London: Routledge & Kegan Paul.

Benacerraf, Paul (1965), 'What Numbers Could Not Be', *Philosophical Review*, 74: 47–73.

Bergson, Henri (1910), *Time and Free Will: An Essay on the Immediate Data of Consciousness*, trans. F. L. Pogson, London: Allen & Unwin.

——— (1911), *Creative Evolution*, trans. Arthur Mitchell, London: Macmillan.

——— (1912), *Matter and Memory*, trans. Nancy Margaret Paul and W. Scott Palmer, London: Allen & Unwin.

Bigelow, John, and Pargetter, Robert (1989), 'Vectors and Change', *British Journal for the Philosophy of Science*, 40: 289–306.

Binkofski, F., and Block, R. A. (1996), 'Accelerated Time after Left Frontal Cortex Lesion', *Neurocase*, 2: 485–93.

Blackburn, Simon (1984), *Spreading the Word: Groundings in the Philosophy of Language*, Oxford: Clarendon Press.

Block, Richard, and Zakay, Dan (2001), 'Retrospective and Prospective Timing: Memory, Attention, and Consciousness', in Hoerl and McCormack (2001), 59–76.

Boghossian, Paul, and Velleman, David (1989), 'Colour as a Secondary Quality', *Mind*, 98: 81–103.

Bourne, Craig (2006), *A Future for Presentism*, Oxford: Clarendon Press.

Braddon-Mitchell, David (2004), 'How Do We Know it is Now Now?' *Analysis*, 64: 199–203.

Broad, C. D. (1923), *Scientific Thought*, London: Routledge & Kegan Paul.

——— (1938), *An Examination of McTaggart's Philosophy*, Cambridge: Cambridge University Press.

Burnyeat, Myles (1990), *The Theaetetus of Plato*, Indianapolis: Hackett Publishing Company.

Butterfield, Jeremy (1984), 'Seeing the Present', *Mind*, 93: 161–76; repr. in Le Poidevin (1998), 61–75.

———— (1999), (ed.), *The Arguments of Time*, Oxford: Oxford University Press.

Byrne, Alex (1993), 'Truth in Fiction: The Story Continued', *Australasian Journal of Philosophy*, 71: 24–35.

Callender, Craig (2002) (ed.), *Time, Reality and Experience*, Cambridge: Cambridge University Press.

Campbell, John (1994), *Past, Space and Self*, Cambridge, Mass.: MIT Press.

Chadwick, Henry (1991) (trans. and ed.), *St Augustine, Confessions*, Oxford: Clarendon Press.

Church, R. M., and Broadbent, H. (1990), 'Alternative Representations of Time, Number, and Rate', *Cognition*, 37: 55–81.

Cloudsley-Thompson, J. L. (1968), 'Time Sense of Animals', in Fraser (1968), 296–311.

Cockburn, David (1997), *Other Times: Philosophical Perspectives on Past, Present and Future*, Cambridge: Cambridge University Press.

Conee, Earl, and Feldman, Richard (1998), 'The Generality Problem for Reliabilism', *Philosophical Studies*, 89: 1–29.

Craig, William Lane (2000), *The Tensed Theory of Time: A Critical Examination*, Dordrecht: Kluwer Academic Publishers.

Crane, Tim (1988), 'The Waterfall Illusion', *Analysis*, 48: 142–7.

Currie, Gregory (1990), *The Nature of Fiction*, Cambridge: Cambridge University Press.

———— (1991), 'Visual Fictions', *Philosophical Quarterly*, 41: 129–43.

———— (1992), 'McTaggart at the Movies', *Philosophy*, 67: 343–55.

———— (1995), *Image and Mind: Film, Philosophy and Cognitive Science*, Cambridge: Cambridge University Press.

———— (1998), 'Tense and Egocentricity in Fiction', in Le Poidevin (1998), 265–83.

———— (1999), 'Can There Be a Literary Philosophy of Time?', in Butterfield (1999), 43–63.

Dainton, Barry (2000), *Stream of Consciousness*, London: Routledge.

———— (2001), *Time and Space*, Chesham: Acumen.

Davidson, Donald (1967), 'Truth and Meaning', *Synthese*, 17: 304–23.

Dennett, Daniel (1991), *Consciousness Explained*, London: Allen Lane.

Dummett, M. (1960), 'A Defence of McTaggart's Proof of the Unreality of Time', *Philosophical Review*, 69: 497–504; repr. in Dummett (1978), 351–7.

———— (1969), 'The Reality of the Past', *Proceedings of the Aristotelian Society*, 69: 239–58; repr. in Dummett (1978), 358–74.

———— (1978), *Truth and Other Enigmas*, London: Duckworth.

Eagleman, D. M., and Sejnowski, T. J. (2000), 'Motion Integration and Postdiction in Visual Awareness', *Science*, 287 (5460): 2036–8.

Evans, Gareth (1982), *The Varieties of Reference*, Oxford: Clarendon Press.

Falk, Arthur (2003), 'Time Plus the Whoosh and Whiz', in Jokic and Smith (2003), 211–50.

Fodor, Jerry (1987), *Psychosemantics*, Cambridge, Mass.: MIT Press.

Forster, E. M. (1927), *Aspects of the Novel*, London: Edward Arnold; repr. in Penguin Classics, ed. Oliver Stallybrass, London: Penguin, 2000.

Fraser, J. T. (1968) (ed.), *The Voices of Time*, London: Penguin.

Friedman, William J. (1990), *About Time: Inventing the Fourth Dimension*, Cambridge, Mass.: MIT Press.

Gale, R. M. (1967) (ed.), *The Philosophy of Time*, London: Macmillan.

——— (1968), *The Language of Time*, London: Routledge & Kegan Paul.

Geach, P. T. (1979), *Truth, Love and Immortality: An Introduction to McTaggart's Philosophy*, London: Hutchinson.

Gettier, Edmund (1963), 'Is Justified True Belief Knowledge?', *Analysis*, 23: 121–3.

Gibbon, J., Church, R. M., and Meck, W. (1984), 'Scalar Timing in Memory', in J. Gibbon and L. Allan (eds.), *Annals of the New York Academy of Sciences*, 423: *Timing and Time Perception*, New York: New York Academy of Sciences, 52–77.

Goldman, Alvin (1967), 'A Causal Theory of Knowing', *Journal of Philosophy*, 64: 357–72.

——— (1976), 'Discrimination and Perceptual Knowledge', *Journal of Philosophy*, 73: 771–91.

Gombrich, E. H. (1960), *Art and Illusion*, Oxford: Phaidon Press.

——— (1964), 'Moment and Movement in Art', *Journal of the Warburg and Courtauld Institutes*, 27: 293–306.

Gregory, R. L. (1966), *Eye and Brain*, London: Weidenfeld & Nicolson.

Grice, H. P. (1957), 'Meaning', *Philosophical Review*, 66: 377–88.

——— (1961), 'The Causal Theory of Perception', *Aristotelian Society*, Suppl. vol. 35: 121–52.

Hale, Bob (1987), *Abstract Objects*, Oxford: Basil Blackwell.

Hamer, Karl C. (1968), 'Experimental Evidence for the Biological Clock', in Fraser (1968), 281–95.

Hestevold, H. Scott (1990), 'Passage and the Presence of Experience', *Philosophy and Phenomenological Research*, 50: 537–52; repr. in Oaklander and Smith (1994), 328–43.

Hirsh, I. J., and Sherrick, J. E. (1961), 'Perceived Order in Different Sense Modalities', *Journal of Experimental Psychology*, 62: 423–32.

Hoerl, Christoph (1998), 'The Perception of Time and the Notion of a Point of View', *European Journal of Philosophy*, 6: 156–71.

——— (1999), 'Memory, Amnesia and the Past', *Mind and Language*, 14: 227–51.

——— and McCormack, Teresa (2001) (eds.), *Time and Memory: Issues in Philosophy and Psychology*, Oxford: Clarendon Press.

Hume, David (1978/1739), *A Treatise of Human Nature*, 2nd edn., ed. P. H. Nidditch, Oxford: Clarendon Press.

Husserl, Edmund (1964/1905), *The Phenomenology of Internal Time-Consciousness*, ed. Martin Heidegger, trans. James S. Churchill, The Hague: Martinus Nijhoff.

Hussey, Edward (1983) (trans. and ed.), *Aristotle's Physics, Book III and IV*, Oxford: Clarendon Press.

Jackson, Frank (1977), *Perception*, Cambridge: Cambridge University Press.

James, William (1890), *The Principles of Psychology*, 2 vols., New York: Henry Holt; repr. as a single volume, Cambridge, Mass.: Harvard University Press, 1983.

Jokic, Aleksandar, and Smith, Quentin (2003) (eds.), *Time, Tense and Reference*, Cambridge, Mass.: MIT Press.

Kant, Immanuel (1929/1787), *Critique of Pure Reason*, trans. Norman Kemp Smith, London: Macmillan.

Kelly, Sean D. (2005), 'The Puzzle of Temporal Experience', in Andrew Brook and Kathleen Atkins (eds.), *Cognition and the Brain: The Philosophy and Neuroscience Movement*, Cambridge: Cambridge University Press, 208–38.

Laird, John (1920), *Study in Realism*, Cambridge: Cambridge University Press.

Lear, Jonathan (1981), 'A Note on Zeno's Arrow', *Phronesis*, 26: 91–104.

Le Poidevin, Robin (1988), 'Time and Truth in Fiction', *British Journal of Aesthetics*, 28: 248–58.

―― (1992), 'On the Acausality of Time, Space, and Space-Time', *Analysis*, 52: 146–54.

―― (1998) (ed.), *Questions of Time and Tense*, Oxford: Clarendon Press.

―― (2003), *Travels in Four Dimensions: The Enigmas of Space and Time*, Oxford: Oxford University Press.

―― and MacBeath, Murray (1993) (eds.), *The Philosophy of Time*, Oxford: Oxford University Press.

Lewis, C. S. (1952), *Mere Christianity*, London: Bles.

Lewis, David (1978), 'Truth in Fiction', *American Philosophical Quarterly*, 15: 37–46; repr. in Lewis (1983*a*), 261–75.

―― (1983*a*), *Philosophical Papers*, i, Oxford: Oxford University Press.

―― (1983*b*), Postscripts to 'Truth in Fiction', in Lewis (1983*a*), 276–80.

Libet, Benjamin (1981), 'The Experimental Evidence for Subjective Referral of a Sensory Experience Backwards in Time: Reply to P. S. Churchland', *Philosophy of Science*, 48: 182–97.

―― (2004), *Mind Time: The Temporal Factor in Consciousness*, Cambridge, Mass.: Harvard University Press.

Locke, Don (1971), *Memory*, London: Macmillan.

Lowe, E. J. (1996), *Subjects of Experience*, Cambridge: Cambridge University Press.

Lucas, J. R. (1973), *A Treatise on Time and Space*, London: Methuen.

MacBeath, Murray (1983), 'Mellor's Emeritus Headache', *Ratio*, 25: 81–8; repr. in Oaklander and Smith (1994), 305–11.

Mackie, J. L. (1977), *Ethics: Inventing Right and Wrong*, Harmondsworth: Penguin.

Mackintosh, N. J. (1983), *Conditioning and Associative Learning*, Oxford: Oxford University Press.

Maclaurin, J., and Dyke, Heather (2002), ' "Thank Goodness That's Over": The Evolutionary Story', *Ratio*, 15: 276–92.

McKay, D. (1958), 'Perceptual Stability for a Stroboscopically Lit Visual Field Containing Self-Luminous Objects', *Nature*, 181: 507–8.

McTaggart, J. E. (1908), 'The Unreality of Time', *Mind*, 17: 457–74.

—— (1927), *The Nature of Existence*, ii, Cambridge: Cambridge University Press.

Markosian, Ned (2002), 'Time', in *Stanford Online Encyclopaedia of Philosophy*, <http://plato.stanford.edu/entries/time/>.

Martin, C. B. and Deutscher, Max (1966), 'Remembering', *Philosophical Review*, 75: 161–96.

Martin, M. G. F. (2001), 'Out of the Past: Episodic Recall as Retained Acquaintance', in Hoerl and McCormack (2001), 257–84.

Maudlin, Tim (2002), 'Remarks on the Passing of Time', *Proceedings of the Aristotelian Society*, 102: 237–52.

Mellor, D. H. (1981), *Real Time*, Cambridge: Cambridge University Press.

—— (1988), 'Crane's Waterfall Illusion', *Analysis*, 48: 147–50.

—— (1995), *The Facts of Causation*, London: Routledge.

—— (1998), *Real Time II*, 2nd edn., London: Routledge.

Moore, A. W. (1997), *Points of View*, Oxford: Clarendon Press.

—— (2001), 'Apperception and the Unreality of Tense', in Hoerl and McCormack (2001), 375–91.

Mozersky, Joshua (2006), 'A Tenseless Account of the Presence of Experience', *Philosophical Studies*, 129: 441–76.

Muybridge, Eadweard (1979/1887), *Animal Locomotion*, New York: Dover.

Nagel, Thomas (1971), 'Brain Bisection and the Unity of Consciousness', *Synthese*, 22: 396–413.

Nerlich, Graham (1994), *What Spacetime Explains*, Cambridge: Cambridge University Press.

Newton-Smith, W. H. (1980), *The Structure of Time*, London: Routledge.

Nijhawan, Romi (1994), 'Motion Extrapolation in Catching', *Nature* 370: 256–7.

Nozick, Robert (1981), *Philosophical Explanations*, Oxford: Oxford University Press.

Oaklander, L. Nathan (1984), *Temporal Relations and Temporal Becoming: A Defence of a Russellian Theory of Time*, Lanham, Md.: University Press of America.

_____ (2002), 'Presentism, Ontology and Temporal Experience', in Callender (2002), 73–90.

_____ (2006), *C.D. Broad's Ontology of Mind*, Frankfurt: Ontos Verlag.

_____ and Smith, Quentin (1994) (eds.), *The New Theory of Time*, New Haven: Yale University Press.

Ornstein, Robert (1972), *The Psychology of Consciousness*, Harmondsworth: Penguin.

Papineau, David (1987), *Reality and Representation*, Oxford: Basil Blackwell.

_____ (1993), *Philosophical Naturalism*, Oxford: Basil Blackwell.

Peacocke, Christopher (1979), *Holistic Explanation: Action, Space, Interpretation*, Oxford: Oxford University Press.

_____ (1987), 'Depiction', *Philosophical Review*, 96: 383–410.

Percival, Philip (1989), 'Indices of Truth and Temporal Propositions', *Philosophical Quarterly*, 39: 190–7.

_____ (1994), 'Absolute Truth', *Proceedings of the Aristotelian Society*, 94: 189–213.

_____ (2002), 'A Presentist's Refutation of Mellor's McTaggart', in Callender (2002), 91–118.

Perry, John (1979), 'The Problem of the Essential Indexical', *Noûs*, 13: 3–21.

Piaget, Jean (1969), *The Child's Conception of Time*, trans. A. J. Pomerans, London: Routledge & Kegan Paul.

_____ and Inhelder, Bärbel (1956), *The Child's Conception of Space*, trans. F. J. Langdon and J. L. Lunzer, London: Routledge & Kegan Paul.

Pine-Coffin, R. S. (1961) (trans. with an introduction), *Saint Augustine, Confessions*, Harmondsworth: Penguin, 1961.

Pöppel, Ernst (1978), 'Time Perception', in Richard Held *et al.* (eds.), *Handbook of Sensory Physiology*, viii: *Perception*, Berlin: Springer-Verlag, 713–29.

Prior, A. N. (1959), 'Thank Goodness That's Over', *Philosophy*, 34: 12–17.

_____ (1967), *Past, Present and Future*, Oxford: Clarendon Press.

Prosser, Simon (2005), 'Cognitive Dynamics and Indexicals', *Mind and Language*, 20: 369–91.

_____ (2006), 'Temporal Metaphysics in Z Land', *Synthese*, 149: 77–96.

_____ (2007), 'Could We Experience the Passage of Time?', *Ratio*, 20 (forthcoming)

Quinton, Anthony (1962), 'Spaces and Times', *Philosophy*, 37: 130–47; repr. in Le Poidevin and MacBeath (1993), 203–20.

Reichenbach, Hans (1958), *The Philosophy of Space and Time*, London: Dover.

Ricoeur, Paul (1984), *Time and Narrative*, i, trans. Kathleen McLaughlin and David Pellauer, Chicago: University of Chicago Press.

_____ (1985), *Time and Narrative*, ii, trans. Kathleen McLaughlin and David Pellauer, Chicago: University of Chicago Press.

_____ (1988), *Time and Narrative*, iii, trans. Kathleen McLaughlin and David Pellauer, Chicago: University of Chicago Press.

Roache, Rebecca (1999), 'Mellor and Dennett on the Perception of Temporal Order', *Philosophical Quarterly*, 195: 231–8.

Robinson, Howard (1994), *Perception*, London: Routledge.

Russell, Bertrand (1903), *The Principles of Mathematics*, Cambridge: Cambridge University Press.

—— (1959/1912), *The Problems of Philosophy*, OPUS edition, Oxford: Oxford University Press.

—— (1921), *The Analysis of Mind*, London: Allen & Unwin.

Schier, Flint (1986), *Deeper into Pictures: An Essay on Pictorial Representation*, Cambridge: Cambridge University Press.

Schlesinger, G. N. (1991), 'E pur si muove', *Philosophical Quarterly*, 41: 427–41.

Schuster, M. M. (1986), 'Is the Flow of Time Subjective?', *Review of Metaphysics*, 39: 695–714.

Shaftesbury, Anthony Ashley Cooper, Earl of (1914/1713), *Second Characters: or the Language of Forms*, ed. Benjamin Rand, Cambridge, Cambridge University Press.

Smith, Quentin (1988), 'The Phenomenology of A-Time', *Dialogos*, 52: 143–53; repr. in Oaklander and Smith (1994), 351–9.

—— (1993), *Language and Time*, New York: Oxford University Press.

Sorabji, Richard (1983), *Time, Creation and the Continuum*, London: Duckworth.

Sperry, R. W. (1964), 'The Great Cerebral Commissure', *Scientific American*, 210: 42.

Steiner, M. (1973), 'Platonism and the Causal Theory of Knowledge', *Journal of Philosophy*, 70: 57–66.

Strawson, P. F. (1974), 'Causation in Perception', in *Freedom and Resentment and Other Essays*, London: Methuen, 66–84.

Swain, Marshall (1979), *Reasons and Knowledge*, Ithaca, NY: Cornell University Press.

Swinburne, Richard (1964–5), 'Times', *Analysis*, 25: 185–91.

—— (1968), *Space and Time*, London: Macmillan; 2nd edn. 1981.

Tallant, Jonathan (2007), 'What is B-time?', *Analysis*, 67 (forthcoming).

Teichmann, Roger (1998), 'Is a Tenseless Language Possible?', *Philosophical Quarterly*, 48: 176–88.

Tooley, Michael (1988), 'In Defense of the Existence of States of Motion', *Philosophical Topics*, 16: 225–54.

—— (1997), *Time, Tense, and Causation*, Oxford: Clarendon Press.

Treisman, Michel (1999), 'The Perception of Time: Philosophical Views and Psychological Evidence', in Butterfield (1999), 217–46.

Tulving, Endel (1983), *Elements of Episodic Memory*, Oxford: Oxford University Press.

Walton, Kendall (1990), *Mimesis as Make-Believe: On the Foundations of the Representational Arts*, Cambridge, Mass.: Harvard University Press.

Wearden, J. H. (2001), 'Internal Clocks and the Representation of Time', in Hoerl and McCormack (2001), 37–58.

Williams, Clifford (1992), 'The Phenomenology of B-Time', *Southern Journal of Philosophy*, 30: 123–37.

Williams, D. C. (1951), 'The Myth of Passage', *Journal of Philosophy*, 48: 457–72; repr. in Oaklander and Smith (1994), 360–72.

Wittgenstein, Ludwig (1953), *Philosophical Investigations*, trans. G. E. M. Anscombe, Oxford: Basil Blackwell.

Woodrow, Herbert (1951), 'Time Perception', in S. S. Stevens (ed.), *Handbook of Experimental Psychology*, New York: John Wiley & Sons, 1224–36.

Wright, Crispin (1980), 'Realism, Truth-Value and Links, Other Minds and the Past', *Ratio*, 22: 112–32; repr. in Wright (1993), 85–106.

—— (1984), 'Anti-Realism, Timeless Truth and *Nineteen Eighty-Four*', in Wright (1993), 176–203.

—— (1993), *Realism, Meaning and Truth*, 2nd edn., Oxford: Basil Blackwell.

Index

Index